Lie Detection and the Law

Torture, Technology and Truth

Andrew Balmer

Routledge
Taylor & Francis Group
LONDON AND NEW YORK

First published 2018
by Routledge

2 Park Square, Milton Park, Abingdon, Oxfordshire OX14 4RN
52 Vanderbilt Avenue, New York, NY 10017

A GlassHouse Book

Routledge is an imprint of the Taylor & Francis Group, an informa business

First issued in paperback 2019

British Library Cataloguing-in-Publication Data
A catalogue record for this book is available from the British Library

Library of Congress Cataloging-in-Publication Data
A catalog record for this book has been requested

ISBN: 978-1-138-85563-2 (hbk)
ISBN: 978-0-367-87250-2 (pbk)

Typeset in Galliard
by Apex CoVantage, LLC

Contents

Author biography

Dr Andrew Balmer is Senior Lecturer in Sociology at the University of Manchester, and is a member of the Morgan Centre for Research into Everyday Lives. He has written three books and from 2018 to 2021 has been editor of *Sociology*, a journal of the British Sociological Association.

Acknowledgements

I am very grateful to numerous people for the help, instruction and advice they have provided over the years of conducting research for this book and turning it into a manuscript. With regard to academic support, I am indebted to the staff and students of the Institute for Science and Society at the University of Nottingham, where I completed my PhD and where most of this work began. I am especially grateful to Prof. Paul Martin, Prof. Ralph Sandland, Prof. Robert Dingwall and Prof. Anne Murcott for their supervision of my work and their invaluable advice. With each piece I write, I find myself stumbling across insights into the practice of being an academic which I would already have gleaned if I had only listened more attentively to them.

Thanks are also due to Prof. Paul Roberts and Prof. Roger Brownsword, who examined my PhD thesis and provided critical feedback on the portions of this book which are adapted from that work. I learnt much from their expert legal perspectives on what I had written. Of course, this being a sociological account, any mistakes regarding the law in this piece are a product solely of my own ignorance.

Additional thanks for many instructive and supportive conversations are due to colleagues in the Department of Sociology, University of Manchester, where this book was written. In particular I would like to thank Dr Wendy Bottero, Dr Gemma Edwards, Dr Vanessa May, Prof. Alice Bloch and Prof. Brian Heaphy. I also benefitted from a period of time at the University of Virginia's Institute for Advanced Studies of Culture, which was supported by a Universitas 21 Visiting Scholarship and where some of the research for this book was conducted.

With regard to assistance in obtaining case materials and sources, I would like to thank Prof. Samuel Gross for supplying a selection of local news articles on the Daniel Gristwood case. I am also very grateful to Donald Connery and Peter Reilly for sharing materials associated with Peter's case. Thank you also to Dr Michael Durrant and Dr Carina Spaulding for pointing me towards relevant literature and providing comments on draft chapters. I am also appreciative for the time spent and information shared by the scientists, lawyers and polygraph operators I interviewed during my PhD, which do not feature in the book but have informed its direction. I am grateful to all of the Routledge team, especially

Prof. John Paterson, Prof. Julian Webb, Collin Perrin and the peer reviewers of the proposal.

Thank you to my family and friends for the many small gifts of time, energy, empathy and patience over the years. My deepest gratitude is always reserved for Dr James Hodgson, my partner, for his extraordinary compassion, generosity and insight.

I alone am responsible for errors or misconstructions in the final text.

Chapter 1

Torture, technology and truth

> I am in the New York State Barracks in Oneida, New York and I am making this statement to Investigator Frank J. Jerome of the New York State Police.[1]

So reads the first line of Daniel Gristwood's signed confession to bludgeoning his wife with a claw hammer.

It was cold in the middle of January 1996, in the town of Clay, New York. The mean temperature was –4°C that month, and the county – along with much of the East Coast of the USA – shivered under three feet of snow. It would become known as one of the great North American blizzards. Christina Gristwood, 26 years old and a mother of five, slept in bed one such freezing night, with her 3-year-old son at her side. And it was in her bed that she was struck three times in the head and face, causing brain damage and partial paralysis. Although she survived after a lengthy surgery, Christina never regained the use of her left side or her memory of the night's events, meaning the only witness able to describe what had happened to her was the assailant.

Earlier that day, Daniel, Christina's husband, went to work in the afternoon and finished late, as was usual for him, around 11pm. During a break around 5 he phoned Christina to check in on the kids and all was well. As his shift came to an end, the Crossroads Tavern beckoned, and Gristwood went drinking with friends until the early hours of Friday morning. When the 29-year-old returned home at 3am he drank a glass of iced tea, smoked a cigarette and watched some television. He would tell the police:

> I went upstairs and into my bedroom and saw my wife laying on the bed and saw that she was naked. I also remember seeing my arm and my fist going toward her head and I remember seeing the hammer on a pillow on the floor next to [her] head. I realize that what I had done up in the bedroom when I could see my arm and fist going toward my wife's head was that I was hitting her in the head. I am very sorry that I did this to my wife because I love her more than anything in the world.[2]

Signed and dated in front of Investigator Jerome, Daniel's ghostly description of watching his arm moving towards Christina's head was the only confession of guilt that he made. The signed document describes how he then ran downstairs to call his aunt to get the kids out of the house, before calling the emergency services to report what seemed to him at the time to have been his wife's suicide. The dispatcher duly sent police and ambulance services to the Gristwood home and the investigation into Christina's assault began.

The polygraph machine, infamously known as 'the lie detector', would become central to his case, since after Daniel volunteered to go to the police station and to be interviewed he was asked to take a test to determine if he was guilty of the assault. His eerie confession was signed at 7.09pm on the Friday, more than fifteen hours after he was brought in for questioning, by which time he had been awake for thirty-four hours, having done a day's work and been out drinking, and after being kept at the police station in a windowless room of about six-by-eight feet, given no food or drink and not allowed to smoke. And, crucially, after an arduous and confusing polygraph exam.

The first of his signed statements was made to Investigator Matthew Tynan, the officer who had been on the scene of the crime and had accompanied Gristwood to the Barracks for his voluntary interview. Daniel signed a typed account of his day, of having gone to work, then out drinking, before returning home to find his wife bloodied in bed. At this point, however, he denied any involvement in the assault and made no mention of seeing his arm going towards her head or of realizing that he had bludgeoned her.

The second statement was made after a further twelve hours of interrogation, this time by Officer Frank Jerome. The new interrogator fast became aggressive, shouting that Gristwood was lying and that he had made inconsistent statements about his actions. But this itself was a lie. Tynan and the senior investigator overseeing the case, Norman Ashbury, would later testify during trial that the officers had absolutely no physical evidence against him and everything that he had said up to that point was consistent and in keeping with the timeline that they had constructed of the evening's events. Christina, the bed, the pillow and the hammer were covered in blood, whilst Daniel was still in the clothes he had worn that night without a spatter on him. So far, he had flat out denied any involvement in the assault and, save for their hunches that his demeanour indicated a hidden callousness, they had nothing to suggest he was guilty. It was a baffling case, for either Gristwood had committed the murder in cold blood, having planned meticulously to leave no evidence connecting him to the murder, or some stranger, without need or motive, had brutally wounded an innocent woman next to her sleeping child. The lie was a gambit – Jerome hoped to wrangle a confession by playing up the case against Daniel and to frighten him into confessing. A confession, they implied, would mean a less severe sentence. But Gristwood would not budge on his account, and with no evidence, Jerome chose to continue with the aggressive interrogation tactics, repeatedly asserting that the police did not believe Daniel and that his failure to confess would only cause him further trouble. But it was still not working, so the officer made another play, telling Daniel that if he wanted to be believed, he should take a polygraph test, implying that if he passed he would be allowed to go home.

Gristwood accepted, and John Fallon, an uncertified operator, was brought in to conduct the examination. Beforehand, Fallon was told that the police did not believe Daniel's account, the implication being that the polygraph might help them prove he was lying. But during the examination Gristwood remained consistent in his story and denied having struck Christina with the hammer. Being sleep-deprived and hung-over, however, he struggled to stay awake. Regardless, Fallon pushed on with the test for four hours, after which he provided Jerome with nothing more than a tentative opinion that the suspect was not telling the entire truth, noting, moreover, that the examination result was fundamentally limited because Gristwood's lack of sleep and recent alcohol consumption could affect the results. In reality, the polygraph exam had failed to provide any useful information, but the subsequent interrogation, based on Jerome's hunch and the examiner's half-baked opinion, led the investigator to draw out Gristwood's second statement, his witnessing of a spectral arm and fist going towards Christina's head and of seeing the hammer on the pillow. Daniel was shown the typed statement and told this is what he had said. He agreed to sign it but immediately returned to claiming his innocence in the face of the signed confession, stating that he had no idea what he had signed and denying once again that he had played any part in the nearly fatal assault on his wife.

On the basis of the confession, the police arrested Gristwood and on 4 March 1996 he was arraigned on an indictment of attempting to murder his wife. Five months later, a jury of three women and nine men were convened to hear the case against him. Amongst the evidence presented were the signed confession, an audio recording of the 911 call he had made and various witness and police testimony regarding his behaviour at the crime scene, during the interrogation and as he made his confession. Prosecuting the case, Senior Assistant District Attorney Nicholas DeMartino encouraged the jury to pay special attention to Gristwood's calm demeanour as he spoke to the emergency services dispatcher. This, he claimed, showed that Daniel Gristwood was not moved by his wife's condition, that it was evidence of his cold disposition and capacity to have committed these ruthless actions. Similarly, Tynan testified that the defendant had showed no emotions during the hours in which he had interviewed him and that he only began to cry once they had told him that they did not believe his story and thought he had tried to kill his wife. The case being made was that Gristwood was a callous man, with little care for Christina, and that he had bludgeoned her with the intent to kill. His demeanour, it was proposed, supported the credibility of the confession that he had signed. The jury found him guilty of second-degree attempted murder and he was sentenced to 12½ to 25 years in prison.

In search of truth

To some extent, many criminal trials can be reduced to the central issue of whether the defendant is telling the truth when he or she pleads not guilty. Daniel Gristwood's case was one such trial. At its heart was a central question: was his denial or his confession a lie? Which should be believed: that he was harangued into providing a false confession after hours of exhaustion or that his mask of innocence momentarily slipped and the police duly seized on his mistake? Since there was limited evidence at the crime scene, much of the trial was concerned with providing testimony on Gristwood's character and behaviour that night, to help the jury to decide whether he was a credible witness or not, and thereby to determine whether he had committed the assault. In effect, the prosecution tried to supply the jury with evidence that Gristwood was a stone-cold liar, whilst the defense tried to convince them he was an innocent man who, quite on the contrary, became the victim of police deceptions. The trial was organised around claims and counter-claims of lying and manipulation.

But how should the legal system judge which of Daniel Gristwood's statements, his confession or his retraction, was the truth? It is a vexing question because it is about more than a single instance of guilt or innocence. It is not just about the truth of Christina's assault, but about how we arrive at that truth. The court must produce justice not only for an individual wronged or an innocent accused, but for society writ large, and in a way which is defensible to the majority (Lynch, 2016). The methods of determining whether someone is lying or not when s/he professes innocence of a crime, especially in cases where other evidence is lacking, exemplify this broader connection between justice and knowledge in society. How lying is managed is situated powerfully within the criminal jury trial with reference to certain conceptions of justice and through certain legal processes of making knowledge.

Legal actors, such as police officers, lawyers and judges, must find ways to make truths about a subject's veracity in a fashion which is practical, workable in a diversity of situations and in a manner which is expedient to their ends and as part of the broader socio-political organisation of justice and knowledge. In pursuit of such ends, the courts have seen the development of rules and procedures for making use of witness testimony, cross-examination and jury decision making. Alongside these changes, courts have often sought the help of scientific experts claiming to provide opinion of use in discerning truth through the fog of human deception, concealment and uncertainty. But direct proof of deception is still lacking, for a method of truly determining whether someone is lying or not has eluded science just as much as it has eluded law. Nonetheless, lie detection practices have long been used to help to judge veracity and thus also guilt in the absence of physical evidence and in ways which are legally binding.

Torture has figured as one such prominent practice within the modes of knowledge production. It has long been central to projects of control within government and law. In some periods religious, cultural and technical forces have brought it into public view, and in others have hidden it away in dungeons and dark rooms, as changing institutional arrangements and political mores make it more or less socially acceptable and rational. The emphasis on the body in torture, bloody and brutal as it is, might lead one to believe that pain was the natural aim of acts such as flaying, burning and quartering. That it was only punishment. But, as so hauntingly elucidated in Foucault's (1995) accounts of shifting forms of power, torture was as much about political ritual as it was about pain (Dreyfus and Rabinow, 2014). In his account of sovereign power, Foucault outlines how the public spectacle of torture, conducted on scaffolds in town centres, served to ritualise the sovereign's ultimate power to destroy life in the face of challenge, for every crime was – to some degree – understood as a challenge to the order which the sovereign guaranteed. The destruction of the criminal's body was not, however, an irrational or animalistic response to the challenge. Rather, argues Foucault:

> Torture rests on a whole quantitative art of pain. But there is more to it: this production of pain is regulated. Torture correlates the type of corporal effect, the quality, the intensity, duration of pain, with the gravity of the crime, the person of the criminal, the rank of his victims. There is a legal code of pain; when it involves torture, punishment does not fall upon the body indiscriminately or equally, it is calculated according to detailed rules.
>
> (Foucault, 1995: 34)

Pain was a form of punishment, clearly, but in its meticulous regulation it also became a kind of art or science of the body, by which a certain kind of truth might be wrested. In other words, torture not only secured the public victory of the sovereign over any challenger, but moved increasingly towards the production of public confessions. The subject was made to confess, to 'validate the truth of the justice of torture and the truth of the accusations. The culmination of the ritual, execution, would also be the culmination of the investigation: truth and power combined' (Dreyfus and Rabinow, 2014: 146).

In important ways, the modern criminal investigation and punishment procedure is quite different.[3] Whereas torture long formed a prominent means through which the truth could be discerned, today we tend to view it as an extreme – something which states resort to only when other means fail. The criminal jury trial has, in significant respects, replaced the public spectacle of torture. Its mechanisms of truth production, taking the form of cross-examination, testimony and so forth, do not resemble torture on the scaffold. Scientific evidence and expert testimony have also to play a role in many jury trials and, through the prevalence of the detective drama genre in mass media culture, it is the forensic trace, the DNA test or fingerprint, which has become the contemporary cultural signifier of guilt (Cole, 2002; Jasanoff, 1998), rather than the strangled, confessing scream. Set against these historical changes, the forensic laboratory seems as far removed from the brutality of the scaffold as one could envision. However, as Latour (2010: 207) argues:

> In order to understand the very special mode of enunciation that one finds in the core of the laboratory, one has to look to torture, to the history of interrogation or the subtle arts of the Inquisition; that is, to the very practices that modern law now regards as shameful and archaic and from which it is at once proud and ashamed to have escaped.

The enunciation of a confession continues to depend upon relations between the body, truth and speech, in the law broadly, but most acutely in the context of criminal evidence. Lie detection technology perhaps more than other means of extracting truth from the body is emblematic of this continued entanglement. With important discontinuities put aside, the forms of lie detection enacted over hundreds of years reflect similar responses to problems of truth. The long history of torture is not severed from our modern modes of resolving conflicts of testimony but is a thread which remains firmly stitched into the socio-political fabric. Putting a spin on Latour's suggestion that our understanding of science, particularly as regards enunciation, must be set against the backdrop of torture, this book explores the development of lie detection methods within torture and the gradual shift towards criminal jury trial and the emergence of the polygraph machine within the criminal justice system as a background to the contemporary management of lie detection within US criminal law.

The (pseudo) science of lie detection

Experiments with scientific methods for recording physical correlates of lying can be roughly traced to Defoe's 1730 essay on the prevention of street crime, in which he writes that 'guilt carries fear always about with it; there is a tremor in the blood of the thief' (quoted in Seagrave, 2004: 3). Defoe advocated holding the wrists and measuring the pulse of a suspected liar to detect his fabrications and secure his veracity. But it is more strictly the 1800s and early 1900s in which lie detection methods properly developed within physiology, psychology and related sciences. Winter (2005, 2012) argues, for example, that the theorisation of the unconscious as a quasi-spatial repository of personal truths made the material memory a focus for biological and psychological study, and subsequently opened up the potential for lie detection technologies. The practitioners of applied psychology, developing their work in the 1870s, created a set of technologies of their own and borrowed from physiological sciences to examine psychic states by monitoring bodily changes. This period saw the invention of numerous lie detection devices, many of which made use of newly invented physiological technologies, such as blood-pressure monitors and tools for recording electrodermal activity. Such experiments were set against a shifting socio-political background of contestation regarding the means we might use to arrive at the truth in a just and rational fashion. However, the developers of lie detection and legal experts never fully resolved the question of how best scientific work might contribute to determining veracity in the absence of material evidence and in the face of contrasting testimony. The result is that a few different lie detection technologies found a home in American life, some being spectacularly successful on their own terms, but never really gaining the kind of scientific and legal standing envisioned by their chief proponents.

Social histories of the polygraph machine, for example, focusing on the first half of the twentieth century, chart the lives of a small group of white men whose influence on the invention and its rise in American life cannot be overestimated. As Alder (2007b), Littlefield (2011) and Bunn (2012) describe, one key set of actors formed around the police chief of Berkeley California, August Vollmer, who campaigned for the professionalisation of the police force and the implementation of scientific means of criminal investigation. He emulated Hugo Münsterberg and William Moulton Marston, two predecessors who had helped bring together psychology and law through their experiments with physiological, forensic methods. But Vollmer himself was no scientist. Instead, he surrounded himself with young male disciples better equipped to experiment with the use of Münsterberg and Marston's methods with the criminal suspects who passed through his police department:

> One disciple, the self-righteous "college-cop" John Larson, had a Ph.D. in physiology from Berkeley with an M.D. from Rush to follow. The other disciple was the man-about-town and entrepreneur Leonarde Keeler, named after Leonardo Da Vinci, though known less grandly as "Nard."
>
> (Alder, 2002: 8)

These men were chiefly responsible for developing both the material form of the polygraph machine and its implementation within questioning and interrogation techniques. Although they developed different combinations of physiological measurement devices, designed to record bodily murmurs as people answered questions about their involvement in or knowledge of a crime, and whilst each went his separate way in this pursuit, by the 1950s the polygraph had taken the form we know from popular media today. It usually comprises a heart rate monitor (to measure pulse), a pneumograph (to measure respiration), an electrical device (to measure perspiration, or 'electrodermal response') and a sphygmomanometer (to measure blood pressure). In the popular imagination, the polygraph still uses a series of automated pens suspended above a rolling sheet of paper to record the readings of those devices; however, in practice today the machine is often connected to a laptop to take the recordings. Indeed, it is now possible to purchase a cheap version of a lie detector machine, complete with several physiological devices and accompanied by software which promises to help anyone determine the truth of someone's claims, and from the comfort of his/her own living room. The 'home lie detector testing kit' claims 80–99 per cent accuracy and requires no expertise.

On its own, however, the polygraph is little more than a collection of tools familiar to us all from doctors' offices. It is the testing procedure adopted which is supposed to turn the machine into a 'lie detector'. The most frequently used technique for questioning subjects is the Control Question Test (CQT). As one of the most prominent experts on lie detection techniques summarises in the *Handbook of Polygraph Testing*:

> The questions are of three general types: (a) Relevant questions . . . "Did you do it?" . . . (b) Control questions – focusing on general, non-specific

> misconducts, of a nature as similar as possible to the issue under investigation (e.g., "Have you ever taken something that did not belong to you?"). (c) Irrelevant questions – focusing on completely neutral issues (e.g., "are you sitting on a chair?") . . . . Typically, the whole question series is repeated three or four times . . . [then there is] a comparison of the responses evoked by the relevant and the control questions. Deceptive individuals are expected to show more pronounced responses to the relevant questions, whereas truthful individuals are expected to show the opposite pattern of responsivity (i.e., more pronounced responses to the control questions).
>
> (Ben-Shakhar, 2002: 4–5)

The alternative approach is not to have a control but rather to look for responses to items involved in a crime that could be known only to those intimately involved with the scene. This practice has been called the Guilty Knowledge Test (GKT) for some time but has recently been rebranded the Concealed Information Test (CIT), which 'utilizes a series of multiple-choice questions, each offering one "relevant" answer (e.g., a feature of the crime under investigation) and several "neutral" (control) answers, chosen so that an innocent suspect would not be able to discriminate them from the relevant one. . . . Typically, if a suspect's physiological responses to the relevant alternative are consistently larger than to the neutral alternatives, knowledge about the event (e.g., crime) is inferred' (Ben-Shakhar and Elaad, 2003: 4).

Such accounts are common in the scientific research on polygraphy and the range of other lie detection methods currently in development, such as the 'layered voice stress test' and the electroencephalograph or 'brain fingerprinting' device. On the surface, at least, such work seems in keeping with scientific research on other forensic techniques, like blood-spatter analysis and ink fingerprinting, each of which emerged in the same period in America. There are further similarities, too, since as with most forensic technologies, the polygraph has been accompanied by a degree of scientific and legal contestation.

Such contestation over forensics often arises by virtue of the social implications of technological innovation (Cole, 2002; Lynch et al., 2008). The long-standing 'social shaping of technology' strand of research in science and technology studies (STS) has well evidenced the way in which 'social factors' are not incidental to technological development and application but are intimately bound-up with such activities: 'the "technical character" of the technology – like what it can do and how it does it – becomes part of the phenomenon to be explained by reference to social and political factors' (Woolgar, 1996: 88). The result of this is a co-production of science, law and social order (Jasanoff, 1995, 2004) in which forensic devices play a crucial role. More recent STS research has sought to evidence the ways in which the material elements of technologies and objects are actors in the formation, maintenance or transformation of social order, political processes and legal practices, resulting in a multiplicity of epistemological and ontological arrangements which do not necessarily sit easily alongside each other (for a review, see Woolgar and Lezaun, 2013).

One outcome of these interrelations is the production of uncertainty. Although scientific work often results in a communal agreement on certain facts, this is often the result of local contingencies and uncertainties being erased in 'global', public accounts (Star, 1985). Moreover, scientific practices not only close down uncertainties but also open them up, manage them, stabilise them and rework them in reflexive moments, but also as part of the everyday production of knowledge, through what Moreira et al. (2009) term 'uncertainty work'. Such work also becomes 'practical' to the extent that it does not disrupt but rather helps maintain the orderliness of scientific research, by preventing otherwise troubling ontological and epistemological uncertainties from interfering with everyday practices (Pickersgill, 2011). In this way, ontological and epistemological uncertainties are not simply problems which must be overcome, for they are regularly managed as part of ongoing socio-technical practices, particularly within periods of 'overflow' in democratic systems (Callon, Lascoumes, & Barthe, 2009) and might even be constitutive, forming part of how objects are enacted and practices performed (Law, 2002; McGoey, 2009, 2012; Pickersgill, 2014; Hollin, 2017b, 2017a).

In this regard, the polygraph fits with a collection of examples interrogated by STS scholars, for uncertainty and contestation have characterised its scientific and legal life. Developed in the 1920s, by the 1960s the machine had spread into a range of legal, governmental and public practices in the United States, but not without controversy. A report on the government's use of lie detection in 1965,[4] for example, claimed: 'There is no "lie detector," neither machine nor human. People have been deceived by a myth that a metal box in the hands of an investigator can detect truth or falsehood.' Some forty years later, in the most thorough review of the research on polygraph methods to date, The National Research Council (part of the National Academies of Science) found the machine was still significantly lacking in validity and reliability. In its opening pages, the report outlined how the Council's investigation had concluded that 'almost a century of research in scientific psychology and physiology provides little basis for the expectation that a polygraph test could have extremely high accuracy' (NRC, 2003: 2). It found that virtually all evidence on polygraphy came from studies of specific-event investigations and that the majority of relevant studies were below the level of quality typically demanded of research funded by the National Science Foundation or the National Institutes of Health. It also highlighted a concern that has been central to the polygraph's history of scientific development, namely its ability to work in the real world when compared with experimental contexts:

> The quality of the studies varies considerably, but falls far short of what is desirable. Laboratory studies suffer from lack of realism, and in the randomized controlled studies focused on specific incidents using mock crimes, the consequences associated with lying or being judged deceptive almost never mirror the seriousness of these actions in real-world settings in which the polygraph is used.
> (NRC, 2003: 3)

Looking across the findings of these two reports, it is clear that in the forty years of scientific research conducted between their publication there was little change in the general scientific consensus on polygraphy. Peculiarly, the first report was published in a period of massive expansion in use of the machine (Alder, 1998), and the second in a period of diversification of its employment (Balmer and Sandland, 2012). Despite the apparent contradiction, between its uncertain status and its continued application, the broad STS approach to the reciprocal enactment of technological, social and political forces still proves instructive here, for it points to the ways in which uncertainties in science might be erased, exposed or used in different socio-legal contexts of the device's application, helping to explain how it is that regular reports on long-standing uncertainties have not augured an end to the polygraph's success.

For example, and perhaps most pressingly, there is an ongoing ontological uncertainty regarding how it is that the intentional action of 'deciding to tell a lie' is connected to measurable changes in the body. In other words, advocates of lie detection continue to have to explain why it is that telling a lie *causes* changes in physiology. In some locations, such as polygraph conferences or police offices, they have success in cementing these connections, largely by communal or strategic assumption, meaning that the question of causation is rarely raised; in others, quite the reverse. The most common account from proponents has remained relatively unrevised from when it was first articulated by the developers of the machine. As Littlefield (2011) shows, these men paired the bodily responses they observed through physiological apparatuses with the notion of a 'deceptive consciousness', forging an ontological alliance between fear and fiction. On this account, it is not lies, as such, that cause bodily changes, but our apprehension that we might be found out. Although generally accepted within the confines of the polygraph research and user communities, this connection is not without problem elsewhere, for it remains contested both in psychological and physiological science more broadly, and in certain spaces in criminal law.

This question of validity, however, tends to fall into the background of polygraph research. The broader question with which the field has been concerned is the epistemological issue regarding how reliably to measure physiological changes to determine when someone has told a lie and to exclude those people who are telling the truth. This question of reducing false positives and negatives continues to be debated and new ways of measuring bodily phenomena, statistically comparing physiological changes, and novel means of questioning subjects, continue to be devised to improve the machine's reliability.

In the realm of polygraph science, then, some things remain relatively constant, whereas others shift more rapidly. This is evidence that some uncertainties are regularly highlighted in the lie detection literature (epistemological ones), whereas others are closed off (ontological ones). The STS concept of 'ontological gerrymandering' is of use here:

> By means of ontological gerrymandering, proponents of definitional explanation place a boundary between assumptions which are to be understood

> as (ostensibly) problematic and those which are not. This "boundary work" creates and sustains the differential susceptibility of phenomena to ontological uncertainty. Some areas are portrayed as ripe for ontological doubt and others portrayed as (at least temporarily) immune to doubt.
>
> (Woolgar and Pawluch, 1985: 216)

In this book, I will show that different socio-legal situations are tied to different ontological sensitivities with regard to the deconstruction of polygraph operations. In some sites, the connection between intention and body is assured, in others it is difficult to maintain. This has also changed over time and continues to shift. I explore this changing ontological sensitivity as part of fluctuating socio-legal practices of managing the polygraph with regard to criminal trials, of using the machine to extract confessions in police interrogations and of adapting it for implementation within sex offender probation programmes.

I also connect this issue to the question which has motivated several social histories of the machine: why hasn't the polygraph failed? As Alder (2007b: 251) reflects: 'Hundreds of investigations of the polygraph have been conducted, only to be reviewed in review essays, studied in metastudies, and analyzed by committees of scientific analysts. Their conclusion? . . . the techniques of lie detection, as used in investigative work by polygraphers, do not pass scientific muster. Yet lie detection lives on.'

Exploring the reason for the lie detector's success in the face of scientific and legal uncertainty, Alder (2007a) claims: 'The lie detector cannot be killed by science, because it is not born of science' and that its true habitat is in news media, film, television, comic books and so forth. I concur with this claim: the polygraph machine has certainly lived a public life as *the* lie detector within the mass media, which has powerfully shaped the way in which it has spread across the United States, informed its gradual assimilation into the United Kingdom and even informs how examinations work. But its role in the courtroom, and its successes, albeit partial and often temporary, cannot be so easily overlooked.

Existing histories (Alder, 2007b; Bunn, 2007; Littlefield, 2011; Weber, 2008) also show that it is the power of confession which has secured a role for lie detection technology within American society. This is convincing and evidences some of the homology between polygraphy and torture. However, there is limited exploration in these accounts of how the polygraph is used, received, articulated and reconfigured in shifting legal winds, alongside the changing socio-political currents so carefully described by these excellent histories. It is by exploring this issue that the overlaps, coherences and contrasts in ontological uncertainty in polygraph science and in its legal status are best understood.

Lying

By contrast with the emphasis placed on the body in lie detection research, lying is generally characterised as a fundamentally social action and has long been conceptualised as a distinctly human phenomenon, a peculiarity of our species' ability to conceal our true thoughts, feelings and intentions from each other (see, for example, Meltzer, 2003). Lying, on this account, is a 'conventionally disfavoured activity that is nevertheless a recurrent feature of social life' (Rodriguez and Ryave, 1990: 196). Despite this characterisation of its social nature, sociology has had surprisingly little to say on the matter. In a compilation of scattered cases of concern, Barnes (1994: 5) concludes: 'Despite its pervasiveness, card-carrying sociologists have done disappointingly little work on lying.' For although deception features as a consideration in several sociological theories (Shilling and Mellor, 2015), it is generally dealt with in passing, such that there remains relatively little sustained engagement with the topic as a *social* rather than individualised, psychological phenomenon.

This is exemplified in the fact that the work of defining lying as a human act has been largely left to philosophers, who – until recently – have generally approached the issue via other concerns, with truth, knowledge, intention, morality, justice and language. The common thread to the more established philosophical investigations has been to define lying as saying something which does not correspond to one's own beliefs or intentions. Many philosophers would add that this is done in an attempt to deceive another person, or in a context in which one warrants or asserts the truth of one's statements.

Plato provides the first, though not nearly the most substantive, discussion of the subject. Most of his thinking on lies and deception plays a secondary role to broader considerations of justice and morality. First, his essay 'Hippias Minor' (see Allen, 1996) deals with lying as part of a dialogue between Hippias and Socrates regarding the question of whether Homer intends to portray Achilles or Odysseus as the better man. Second, *The Republic* (see Bloom, 1991) considers the possible virtue of a 'noble lie' in the founding of a just society. Though the treatment of the noble lie in *The Republic* is better known, it is in the earlier piece, the 'Hippias Minor', that Plato gets to grips with the concept of lying and whether it is ever morally correct to lie.

Hippias, a boastful and sometimes preposterous polymath, enters into dialogue with Socrates by putting forward a claim that Achilles is the better man than Odysseus, who is the more wily. Although on the face of it a literary discussion, the issue relates to a broader debate in the period regarding the nature of virtue. Hippias, a famous Sophist, believed virtue could be taught because it was an art. Plato disagreed, arguing that virtue cannot be an art.

The claim regarding Achilles is received by Socrates in a manner of quite obviously feigned humility, as he professes not to understand Hippias' arguments, showing-up the self-aggrandising behaviour of his conversational partner, but also indicating for the reader that Socrates is not the naïve listener he appears to be playing. From the start, Hippias is on the back foot and is quickly out of his depth, whereas Socrates seems to be concealing his true motives, cannily leading Hippias towards a paradoxical conclusion. Although Socrates manipulates Hippias towards an unsustainable conclusion in this dialogue, it is the Sophists such as Hippias who are remembered for being deceptive, for their willingness to provide education in virtue only to those able to pay for it. Since the dialogue will be concerned with whether ignorance or deception signifies the better man, it is quite possible that Plato is having fun with the subject matter and the Sophistic method, through the staging of the dialogue: the reader is in on the fact that this discussion of deception is leading Hippias into a trap.

Socrates proceeds through the dialogue in an unrelenting critique of Hippias' contention that Achilles is the better man. Hippias quotes Achilles from Homer's epic poem, the *Iliad*, in which he says:

> Son of Laertes, sprung from heaven, crafty Odysseus, I will speak out plainly the word which I intend to carry out in act, and which will, I believe, be accomplished. For I hate him like the gates of death who thinks one thing and says another. But I will speak that which shall be accomplished.
>
> (Allen, 1996: 32 – Hippias Quotes the Illiad Book XI: modification of lines 308–14)

Hippias delivers this as evidence that Achilles is simple and true and therefore the better man, whereas Odysseus is wily and false, and therefore the less virtuous. The Iliad quotation exemplifies the moral issue with which the Minor is concerned, namely the question of whether it is better to do wrong voluntarily or in ignorance. Socrates rains down a deluge of examples upon the embattled Hippias, who is trying to maintain that Achilles is better, but with each example and each seemingly natural conclusion Hippias finds himself losing the argument. This leads the dialogue into a curious position. The conclusion is apparently an impasse, for it seems to support the idea that if someone speaks falsely intentionally by having knowledge of the truth (i.e. they tell a lie), then they are a better person than someone who speaks falsely in ignorance (i.e. they make a mistake) because only those who know the truth can tell it, whereas those in ignorance are doomed to repeat their mistakes. Hippias won't stand for this absurd conclusion but is also unable to show why it is wrong. To see why Plato leads us to this impasse we must understand what it is that the argument is really trying to achieve. Allen (1996: 29) summarises the key result of the paradox regarding virtue and art (techne):

> If virtue is an art, then the good man can voluntarily do wrong, and it is better to do wrong intentionally than unintentionally [i.e. it is better to lie knowingly than to be in error]. If this is not true, then virtue is not an art.

Plato's not so gentle boxing of Hippias leaves him with a conclusion which he cannot tolerate but which is the result of his own beliefs. Either he accepts the seemingly absurd idea that intentional lying is better than error, or accepts that virtue is not an art. But neither appeals to him. Exploring this account shows that lying is implicitly wrong in both Hippias' and Plato's positions and that it is the moral status of the lie which is used to examine the broader ontological question of what virtue is, an art or something else entirely. For Plato, the resolution is easier: virtue cannot be an art, for otherwise it leads to this absurd outcome.

A brief look at Plato's thoughts on lying also evidences the complex ways in which lies are tied up with questions of knowledge, morality, justice and law in the history of philosophy. It marks a formative distinction between truth, lying and error and sets a precedent of philosophical argument about its moral status. Lying, in almost every account since, will be treated as immoral by nature. In all future definitions and moral exegesis, there is also some consideration of the intention of the speaker and his/her knowledge of the world, just as with Achilles and Odysseus in the Minor. Whether in Aristotle, Aquinas, St. Augustine, Rousseau or Kant, the question of lying is thereby treated as a question regarding the opacity of will, for it is this which makes possible the moral defect in the lie: that it is used to deceive.[5]

The most substantive discussion of lying which we have in the canon is from Saint Augustine (see Schaff, 1887: 457), who says: 'There is a great question about lying,' in the opening line of his treatise on the matter. It is a question regarding whether it is ever right to tell a lie, even a well-meaning, charitable one. The first step in his answer is to determine what a lie is. But, he discovers, in pursuit of it one will find the trail 'very full of dark corners' and that it 'hath many cavern-like windings, whereby it oft eludes the eagerness of the seeker; so that at one moment what was found seems to slip out of one's hands, and anon comes to light again, and then is once more lost to sight' (Schaff, 1887: 457). Despite getting lost as he explores the questions of what a lie is (he never quite resolves the key issue he raises regarding the relation between lying and deception), his analysis is skilful and remains the more prominent of these two early works. Of the two key essays, 'De Mendacio' (On Lying, 395AD) and 'Contra Mendacio' (Against Lying, ~400AD), the former is more instructive.

Inheriting Plato's concern with differentiating lying and error, Augustine clarifies that it cannot be the case that a lie is merely any false statement that we make, because we make a great many utterances that are false with regard to the reality of affairs but which we earnestly and innocently believe to be true. The correlation between the statement's claim and the fact of the world is therefore irrelevant, rather 'from the sense of his own mind, not from the verity or falsity of the things themselves, is he to be judged to lie or not to lie' (Schaff, 1887: 458).

Instead, Augustine argues that lying is fundamentally characterised by 'duplicity' – a doubling of thought:

> the heart of him who lies is said to be double; that is, there is a double thought: the one, of that thing which he either knows or thinks to be true and does not produce; the other, of that thing which he produces instead thereof, knowing or thinking it to be false.
>
> (Schaff, 1887: 458)

Based on his interpretation of the Christian Bible, he argues that humans are made in God's image, and since God is only truthful, human nature is also to be truthful. On this account, when we think of anything, the first thought we have is always our true belief on the matter: it is the default or 'pre-potent' response to the world. This does not mean that we always know the factual truth, because our beliefs might be in error, but we always know what we, at least, truly believe. Developing this logic, Augustine argues that to create a lie we must create a second thought, one which is contrary to our true belief and which therefore produces duplicity of thought. Griffiths (2004: 25) succinctly articulates Augustine's ontological position:

> The characteristic mark of the lie is duplicity, a fissure between thought and utterance that is clearly evident to the speaker as she speaks. Lying words are spoken precisely with the intent to create such a fissure: the liar takes control of her speech, and marks it as her own when she separates it from her thought and grants it autonomy. When this is done, speech becomes the possession and instrument of the speaker. When it is not done, speech comes to birth spontaneously, as an act of gratitude and praise to its giver. Lies require effort; truth none.

To speak anything but truth is therefore a movement away from God and towards sin. Lying, says Augustine, comes from the devil, who is the father of lies.[6] As such, Christians should never tell a lie; although it might regularly seem that this is the better thing to do, Augustine argues that there is no Christian justification for lying. Augustine carefully outlines the importance for Christians of not lying. However, at the same time, he shows that it is difficult to know where the truth ends and a lie begins. Augustine explores the case of a cynical acquaintance, who does not believe you, no matter what you say. Perhaps, we can speculate, this is because of your condemnable propensity for lying, and your friend is right not to trust you. Or perhaps the friend is to blame for being too distrustful. Whatever the cause, the outcome is that you will struggle to convince him of the truth of your beliefs. Augustine imagines a scenario in which you meet the cynical acquaintance at a crossroads, in an area familiar to you but unknown to the other, who is lost and needs your help to get home. One road leads to safety (road S), the other to certain assault at the hands of muggers (road M). If your acquaintance only trusted you, you could safely send him home via road S by simply telling him the truth. But he does not, so you cannot. Instead, you must lie to convince him of the truth. You tell him that road M is the safe route home and he does not believe you; he goes via road S and finds himself returned safely to domestic routine. If, on the other hand, you wish to punish him for never trusting you, then you can tell him the truth, that road S is the safe route home, which he will not believe and will therefore find himself mugged as he discounts your advice and takes road M. Through these imagined cases, Augustine shows you can tell the truth in order to deceive, and lie in order to communicate the truth. But which of these is a lie, or is it neither? The problem of defining the role of deception in lying persists, as philosophers continue to debate whether or not the definition of lying requires an intention to deceive another person. The other person remains tangled in a quest to determine why it is that lying is immoral, and to thereby be able to once and for all distinguish truth from lying, bad intentions from good and be able to educate civilians in the proper route to virtue.

Although the past two millennia have provided these and other canonical explorations, it is the past two decades which have seen an expansion of philosophical interest in lies, deception and related concepts, such as bullshit and malingering. Contemporary philosophy, nonetheless, continues in the same tradition of Plato and Augustine, developing conceptual schemas which demarcate lies from other actions, such as joking and bluffing in ever greater refinement (Fallis, 2009). They do so by testing their definitions against fictional cases, much as Plato did in using the actions of the heroes in Homer's epic, or as Augustine did in his exploration of the mistrustful acquaintance. Such cases are predominantly of philosophers' creation and are designed primarily to show up differences in competing definitions, help rework them and judge one to be more successful than another.

But these and other philosophical expositions, whilst rich and rewarding in their own right, do not help in determining how it is that we do, or should, judge someone to be lying or not in a practical, everyday fashion. However sophisticated a definition philosophers might devise, people will still be left to judge cases on a daily basis where intentions are hidden from us, knowledge uncertain and memory corrupt. Moreover, what the continued work on philosophy shows is that central uncertainties identified by Plato and Augustine remain. Philosophers too have a problem with the 'real world' application of their definitions, which often fail to mirror how the word 'lie' is used in everyday life. For all its exemplary analysis in demarcation, the result of two thousand years of philosophy has really been to confirm that the lie itself is an uncertain object.

Despite its relative silence, sociology has a role to play here, by providing the groundwork for a theoretically informed, empirical approach to the study of lying in social life. First, German sociologist and philosopher Georg Simmel provides a fractured and difficult account of the importance of secrets (and to a lesser extent, lies) in social life, but one which remains useful to such ends. Rather than being founded on complete truthfulness, society, he argues, is instead cemented on ever shifting sands of selective disclosure, veiled truths, secrecy and, yes, outright lies. As Schermer and Jary (2013: 129) summarise: 'Sharing fully with others one's inner-flow of consciousness is impossible, and selection also provides scope for purposefulness. Sincere self-disclosure and its opposite only come about in the context of a wider, ever-present, 'not knowing' about one another.'

To add to the issue of the pervasiveness of not-knowing, there is also a problem of the situated nature of lying. In one of the few sociological papers developing a more empirical approach, Hunt and Manning (1991: 52) argue:

> Lying is not an obvious matter: it is always socially and contextually defined with reference to what an audience will credit; thus its meaning changes and its effects are often ambiguous. [. . .] In a sense, lies do not exist in the abstract; rather they are objects within a negotiated order.

I agree that definitions which presuppose a singular object of 'lying' fail to understand the ways in which lies become salient within different sets of practices, falling into the philosophical trap of trying to find a universal way of determining why lies are wrong and what they comprise.

It is not only the negotiations which result from differences in place which provide trouble for a general account of lying – there is also a question of time and change. Simmel (1950) is concerned with how the social forms of secrecy and lying might change alongside other social arrangements. In pre-modern periods, he suggests, secrets and lies were part of the social fabric but did not occur as frequently as they do today:

> In a small and narrow circle, the formation and preservation of secrets is made difficult even on technical grounds: everybody is too close to everybody else and his circumstances, and frequency and intimacy of contact involve too many temptations of revelation.
>
> (Simmel, 1950: 335)

As societies grow larger and become more urbanised, a shift in intimacy and differentiation therefore occurs, through which we become less familiar with and less similar to each other. Simmel argues that the result is a vast increase in the amount of information about ourselves, our desires and thoughts, which we must learn to keep to ourselves. A further effect is that lies become more dangerous the more differentiated and individualised we become, 'something much more devastating than it was earlier, something which questions the very foundations of our life' (Simmel, 1950: 313).

Simmel's generalisation of pre-modern and modern periods leads us to the right concern but is too sweeping. Lies might have different effects and can be understood differently over time as the valence of truth, trust, risk and faith also shift. At the same time, lies can still be seen as a mundane fact of social existence, one simply to be expected, even as they periodically take centre stage in the political drama of truth and democracy. Take, for example, the publication of the 'Pentagon Papers', a set of government documents that revealed the extent of deception and misinformation propagated by all levels of the US government during the Vietnam War. Such a significant socio-political event in the history of mass deception might prompt concerns that something had changed in politics which needed to be addressed. But as Hannah Arendt (1972: 4–5) wrote in her incisive response to the publication:

> Truthfulness has never been counted among the political virtues, and lies have always been regarded as justifiable tools in political dealings. Whoever reflects on these matters can only be surprised by how little attention has been paid, in our tradition of philosophical and political thought, to their significance.

Arendt is justified in her expression of surprise at the lack of scholarship on lying and deception as human actions, given their significance for the organisation of social relations, not only in the day-to-day experience of everyday people, but in the higher echelons of government. The more recent drama of President Trump's election and the decision of the UK population to 'Brexit' from the European Union are only two contemporary events in a longer, cluttered history of political crises regarding lying and deception. Trump quickly polarised opinion, with some seeing him as a vanguard for a new political authenticity, whereas others saw the rise of a masterful bullshitter. The Brexit campaigns were similarly steeped in discussions over the truthfulness of politicians, with the infamous red bus and its promise of more money for the NHS becoming, for many, a lasting symbol of the perils of political spin, and of public credulity. The pervasiveness of lies and worries over their prevalence seem to warrant serious consideration just as much as they seem to make such lies slip into the background. We need a sociological approach which takes account of the ways in which lying might be understood, managed or negotiated differently across contexts and over time.

Where science and law are entangled, the question of truth is often at stake, and lies and their detection often figure as a part of these concerns. As such, an examination of how lie detection is conducted within different socio-legal contexts offers the potential to better understand not only how technologies of truth have been positioned with regard to law, justice and social order, but also with how lying itself figures in such different situations. In this book, I cannot develop a substantive sociological approach to lies as social actions writ large, but I do seek to support such a broader sociological programme by way of a consideration of some of the ways in which lying is articulated in some of the different socio-legal practices in which torture and lie detection technologies have figured, and thereby of how lying is understood as a part of social organisation more broadly.

Overview of the book

Chapter 2 explores the use of torture as a form of lie detection in practices of 'trial by ordeal'. Against this background, it explains the resurgence in England of state-sanctioned torture as a form of interrogation in the Elizabethan period before outlining the literature on the emergence of the polygraph machine and the trial by jury in the United States. Together, these threads depict the shifting conceptualisations of the body and its relation to truth, setting up themes which recur throughout the text.

Chapter 3 provides an overview of the tripartite heuristic used to explain the legal status of the polygraph machine in US criminal courts, that it is either: inadmissible; admissible with prior-stipulation; or admissible without prior-stipulation. It also argues for greater nuance in characterising the socio-legal status of the polygraph, which is followed up in Chapters 4 and 5.

Chapter 4 begins to provide a more detailed picture of how the polygraph machine is challenged and managed within US state supreme courts, outlining the 'exclusionary toolkit' and how this is used to highlight uncertainties in lie detection science and operation.

Chapter 5 extends this analysis by examining how ontological uncertainties in polygraph research and practices are negotiated in legal settings by reference to broader techno-political currents in the United States. It shows that the ontological connections between body, truth and lying are enacted differently over time in a case study of the Massachusetts state supreme court's cases involving polygraph admissibility.

Chapter 6 explains how the polygraph machine is used in police interrogations of criminal suspects and witnesses. It describes in detail the deceptive techniques used by polygraph interrogators to manipulate subjects into confession. The chapter evaluates these techniques in light of the tragic case of Peter Reilly, showing how lie detection practices can bring about false confessions.

Chapter 7 explores the use of the polygraph at the socio-legal periphery of the criminal trial, namely in probation and treatment programmes for sex offenders. It describes the emergence of sex offender polygraph testing in the USA and the UK, setting this against a background of moral panics in both countries, but also linking it to a longer history of socio-technical practices through which certain groups have been made into suspect subjects. This chapter critically assesses the confessional technique used in the management of sex offenders, detailing the way in which power is structured in this most recent and significant shift in the polygraph's history.

Chapter 8 reviews the ways in which uncertainty in lie detection has figured in its techno-legal configurations within socio-legal situations. It explores the implications of this account for a sociological approach to lying and concludes the story of Daniel Gristwood's confession to bludgeoning his wife.

Notes

1 Quoted from Daniel Gristwood's confession, page 1. For the case history: *Gristwood v State of New York*, 39 Misc 3d 1226[A], 971 NYS2d 71, 2013 NY Slip Op 50737[U], 2013 N.Y. Misc. LEXIS 1977 (2013).
2 Quoted from Daniel Gristwood's confession, page 2.
3 Even in the United States, the main geographical focus of this book, where in some states punishment by execution remains.
4 *Use of Polygraphs as "Lie Detectors" by the Federal Government*, H. Rep. No. 198, 89th Cong. 1st Sess., at p. 1.
5 Though it is worth noting that for Augustine the deeper moral defect of the lie is that it moves us away from God, and for Kant it is that lies erode the trust on which society is founded.
6 Bible, John 8: 44–45: 'When he speaks the lie he speaks by expropriation, because he is a liar and father of the lie.'

References

Alder, K. (1998) To Tell the Truth: The Polygraph Exam and the Marketing of American Expertise, *Historical Reflections* 24 (3): 487–525.

Alder, K. (2002) A Social History of Untruth: Lie Detection and Trust in Twentieth-Century America, *Representations* 80 (1): 1–33.

Alder, K. (2007a) America's Two Gadgets: Of Bombs and Polygraphs, *Isis* 98 (1): 124–137.
Alder, K. (2007b) *The Lie Detectors: The History of an American Obsession*, New York: Free Press.
Allen, R.E. (1996) *The Dialogues of Plato, Volume 3: Ion, Hippias Minor, Laches, Protagoras*, London: Yale University Press.
Arendt, H. (1972) *Crises of the Republic*, London: Harcourt, Brace and Company.
Balmer, A.S. and Sandland, R. (2012) Making Monsters: The Polygraph, the Plethysmograph and Other Practices for the Performance of Abnormal Sexuality, *Journal of Law and Society* 39 (4): 593–615.
Barnes, J.A. (1994) *A Pack of Lies: Towards a Sociology of Lying*, Cambridge: Cambridge University Press.
Ben-Shakhar, G. (2002) A Critical Review of the Control Question Test (CQT), pp. 103–126 in M. Kleiner (ed.) *Handbook of Polygraph Testing*, London: Academic Press.
Ben-Shakhar, G. and Elaad, E. (2003) The Validity of Psychophysiological Detection of Information with the Guilty Knowledge Test: A Meta-Analytic Review, *Journal of Applied Psychology* 88 (1): 131–151.
Bloom, A. (1991) *The Republic of Plato* (2nd Edition), New York: Basic Books.
Bunn, G.C. (2007) Spectacular Science: The Lie Detector's Ambivalent Powers, *History of Psychology* 10 (2): 156–178.
Bunn, G.C. (2012) *The Truth Machine: A Social History of the Lie Detector*, Baltimore, MD: Johns Hopkins University Press.
Callon, M., Lascoumes, P. and Barthe, Y. (2009) *Acting in an Uncertain World: An Essay on Technical Democracy* (translated by Graham Burchell, originally published in 2001), Cambridge, MA: MIT Press.
Cole, S. (2002) *Suspect Identities: A History of Fingerprinting and Criminal Identification*, Boston, MA: Harvard University Press.
Dreyfus, H.L. and Rabinow, P. (2014) *Michel Foucault: Beyond Structuralism and Hermeneutics*, Chicago, IL: University of Chicago Press.
Fallis, D. (2009) What Is Lying?, *Journal of Philosophy* 106 (1): 29–56.
Foucault, M. (1995) *Discipline and Punish: The Birth of the Prison*, New York: Random House.
Griffiths, P.J. (2004) *Lying: An Augustinian Theology of Duplicity*, Grand Rapids, MI: Brazos Press.
Hollin, G. (2017a) Autistic Heterogeneity: Linking Uncertainties and Indeterminacies, *Science as Culture* 26 (2): 209–231.
Hollin, G. (2017b) Failing, Hacking, Passing: Autism, Entanglement, and the Ethics of Transformation, *BioSocieties*, https://doi.org/10.1057/s41292-017-0054-3.
Hunt, J. and Manning, P. (1991) The Social Context of Police Lying, *Symbolic Interaction* 14 (1): 51–70.
Jasanoff, S. (1995) *Science at the Bar: Law, Science, and Technology in America*, Cambridge, MA: Harvard University Press.
Jasanoff, S. (1998) The Eye of Everyman: Witnessing DNA in the Simpson Trial, *Social Studies of Science* 28 (5/6): 713–740.
Jasanoff, S. (2004) The Idiom of Co-Production, pp. 1–12 in S. Jasanoff (ed.) *States of Knowledge: The Co-Production of Science and Social Order*, New York, NY: Routledge.

Latour, B. (2010) *The Making of Law: An Ethnography of the Conseil D'état*, Cambridge: Polity Press.
Law, J. (2002) On Hidden Heterogeneities: Complexity, Formalism, and Aircraft Design, pp. 116–141 in J. Law and A. Mol (eds.) *Complexities: Social Studies of Knowledge Practices*, Durham, NC: Duke University Press.
Littlefield, M.M. (2011) *The Lying Brain: Lie Detection in Science and Science Fiction*, Ann Arbor: University of Michigan Press.
Lynch, M. (2016) Preliminary Notes on Judges' Work: The Judge as a Constituent of Courtroom "Hearings", pp. 99–130 in M. Travers and J.F. Manzo (eds.) *Law in Action: Ethnomethodological and Conversational Analytic Approaches to Law*, Abingdon: Routledge.
Lynch, M., Cole, S.A., Mcnally, R. and Jordan, K. (2008) *Truth Machine: The Contentious History of DNA Fingerprinting*, Chicago, IL: University of Chicago Press.
McGoey, L. (2009) Pharmaceutical Controversies and the Performative Value of Uncertainty, *Science as Culture* 18 (2): 151–164.
McGoey, L. (2012) The Logic of Strategic Ignorance, *The British Journal of Sociology* 63 (3): 533–576.
Meltzer, B. (2003) Lying: Deception in Human Affairs, *International Journal of Sociology and Social Policy* 23 (6/7): 61–79.
Moreira, T., May, C. and Bond, J. (2009) Regulatory Objectivity in Action: Mild Cognitive Impairment and the Collective Production of Uncertainty, *Social Studies of Science* 39 (5): 665–690.
National Research Council (2003) *The Polygraph and Lie Detection*, Washington, DC: National Academies Press.
Pickersgill, M. (2011) Ordering Disorder: Knowledge Production and Uncertainty in Neuroscience Research, *Science as Culture* 20 (1): 71–87.
Pickersgill, M. (2014) The Endurance of Uncertainty: Antisociality and Ontological Anarchy in British Psychiatry, 1950–2010, *Science in Context* 27 (1): 143–175.
Rodriguez, N. and Ryave, A. (1990) Telling Lies in Everyday Life: Motivational and Organizational Consequences of Sequential Preferences, *Qualitative Sociology* 13 (3): 195–210.
Schaff, P. (1887) *A Select Library of the Nicene and Post-Nicene Fathers of the Christian Church, Volume 1: The Confessions and Letters of St. Augustin*, Buffalo, NY: The Christian Literature Company.
Schermer, H. and Jary, D. (2013) *Form and Dialectic in Georg Simmel's Sociology: A New Interpretation*, Basingstoke: Palgrave Macmillan.
Seagrave, K. (2004) *Lie Detectors: A Social History*, London: MacFarland and Company.
Shilling, C. and Mellor, P.A. (2015) For a Sociology of Deceit: Doubled Identities, Interested Actions and Situational Logics of Opportunity, *Sociology* 49 (4): 607–623.
Simmel, G. (1950) *The Sociology of Georg Simmel*, New York, NY: Free Press.
Star, S.L. (1985) Scientific Work and Uncertainty, *Social Studies of Science* 15 (3): 391–427.
Weber, S. (2008) The Hidden Truth: A Sociological History of Lie Detection (PhD Thesis), London: London School of Economics.
Winter, A. (2005) The Making of "Truth Serum", *Bulletin of the History of Medicine* 79 (3): 500–533.

Winter, A. (2012) *Memory: Fragments of a Modern History*, Chicago, IL: University of Chicago Press.
Woolgar, S. (1996) Technologies as Cultural Artefacts, pp. 87–102 in W.H. Dutton and M. Peltu (eds.) *Information and Communication Technologies: Visions and Realities*, Oxford: Oxford University Press.
Woolgar, S. and Lezaun, J. (2013) The Wrong Bin Bag: A Turn to Ontology in Science and Technology Studies? *Social Studies of Science* 43 (3): 321–340.
Woolgar, S. and Pawluch, D. (1985) Ontological Gerrymandering: The Anatomy of Social Problems Explanations, *Social Problems* 32 (3): 214–227.

Chapter 2

Truth and lies from torture to technology

> As old as the history of crime is the history of cruelties exercised, in the service of justice, for the discovery of criminal facts. Man has the power to hide his knowledge and his memories by silence and by lies, and the infliction of physical and mental pain has always seemed the quickest way to untie the tongue and to force the confession of truth.
>
> (Münsterberg, 1907: 614)

When Hugo Münsterberg, Professor of Psychology at Harvard University, wrote these words over a hundred years ago, he sought to condemn the use of torture to extract confessions during police interrogations, a practice known as the 'third degree'. He believed the nascent science of psychology would prove to be a 'magnifying-glass for the most subtle mental mechanism, and by it the secrets of the criminal mind may be unveiled' (Münsterberg, 1907: 622). Such a study of the mind promised to civilise society not only by uncovering criminality but also by making justice more humane.

The claims of psychologists in the early 1900s linked science and law in an explicit fashion, proposing not simply an epistemological framework for the study of human behaviour but also a politics and an ethics to govern our actions. Not for the first time, scientists were working to reconfigure the co-production of knowledge and social order (Jasanoff, 2004). In the decades that followed, psychology gradually found its second home outside of the university, and became commonplace within the courtrooms, police stations and forensic laboratories of the criminal justice system. This was not only an argument for an ideological shift, however, but a deeply practical one, in which a range of techniques and technical devices were developed, tested and put to use in the field. The polygraph machine, the criminal psychologists' chief tool of deception detection, was just one amongst several such techniques, but a crucial one nonetheless, for it played a key role in the development of the legal framework for using scientific evidence during criminal trials in the United States, and it would go on to become one of the most widely used technologies in American policing practices.

Today, neuroscience has supplanted psychology in the quest for a civilising science, one that might displace physical abuse in the investigation of crimes and better expose the corrupt, incorrigible and deviant within our midst. Brain imaging technologies especially have become the focus of research aimed at finally uncovering the secrets of criminal intent. Torture today is less visible within the police interrogation room, so it is in the context of averting terrorism that many of these neuroscience experiments find their social justifications (Littlefield, 2009). Alongside this shift there has been a concurrent change in the epistemological claims made regarding the criminal mind, but the political expediency remains much the same. Daniel Langleben, for example, one of the inventors of contemporary brain-scanning methods for lie detection, claims:

> Any kind of crime, including terrorism, happens in the brain. Terrorism and crime happen here first, that's undeniable. And if we can keep it there, that would be nice. It would be good I think even for the perpetrator of the crime to have not done it, or in some way to not bring it to the catastrophic proportions of a mature plan. More than that, the investigative techniques available for people today are terrible. People are either resorting to torture, which is just out of the question, or resorting to legal questions that don't bring them [answers] fast enough. Forensic neuroscience is going to happen.
>
> (Langleben quoted in Hyland, 2005)

Langleben's comments are fitting for the politics of the day. We no longer hunt for witches or depend upon divine intervention to deal out justice. Terrorists are the witches of our age and science the location of benevolent omniscience. It may seem a natural and important progression in understanding that the brain should now be the place in which we expect to discern the truth, since neuroscience has begun to permeate much of Western governance (Choudhury and Slaby, 2011; Pickersgill and Van Keulen, 2011). However modern it may seem, justifying research into brain-scanning lie detection with the promise to end torture is contiguous with past justifications. Physical duress has often provided the backdrop to the development, marketing and administration of technological forms of lie detection (Littlefield, 2009). Torture has been used for thousands of years within the practices of justice and the broader construction of social order. Whilst on the face of it torture is itself a form of lie detection, apparently used to determine the truth of someone's thoughts and actions, memories and intentions, in certain respects it is also the practice in which are expressed the limits of an era's epistemic apparatus for determining the truth. It is thus often a practice in relation to which new ontologies of truth and fiction are established as existing measures fail. The development of functional MRI lie detection is one such example.

Torture, truth and lying have thereby long been entangled. A cursory look at the history of lie detection and torture practices might also suggest a narrative of enlightenment, a gradual improvement in our civility, journeying from the bloodied scaffold of medieval public spectacles to the pristine white surfaces of the brain scanner. The view is ostensibly of a passage from savagery to science. However, the route is travelled in both directions. What once was considered abhorrent becomes justified again; what was once routine falls into disrepute. Whilst the torture scaffold appears to be a relic of the past, the contemporary uses of waterboarding, hooding and deprivation (Gross, 2010) force us to confront the continued legacy of state-sanctioned torment not only in the West but around the world. Nor is the history of torture one of a transition from irrationality to sense. Medieval torture was conducted strategically, as part of shifting practices of knowledge production, law and government. Moreover, the conceptual apparatus put to work in making sense of torture in medieval England is not entirely divorced from that which we use today. Torture and science, confession and revelation were and remain deeply entangled in efforts to govern ourselves.

Torture and the criminal trial

In England, the use of torture in criminal trials was officially banned in 1215 when the Fourth Lateran Council forbade clerical participation in the trial by ordeal (Groot, 1982). Trials of ordeal involved suspects being subjected to various kinds of physical torture, the outcomes of which were used to determine the guilt or innocence of the accused. The church's role in the trial was more than symbolic. It is true that its involvement was understood to allow for God's perfect insight to be heard, so that the resultant punishments or exonerations were divinely sanctioned, but it also had a practical role. In the administration of torture, it was often the clergy who were directly responsible for judging whether its physical effects and the body's recovery were indicative of guilt or innocence (Pilarczyk, 1996). As such, the loss of a divinely endorsed sign of culpability in 1215 left the English justice system not simply with a philosophical but also with a practical problem, opening a space in which new criminal procedures could develop, one that the jury as we know it today would eventually come to inhabit (Fisher, 1997).

There is some disagreement in the literature about how long torture, in the form of the ordeal, had been used prior to 1215. Kerr et al. (1992) argue that it was only for around 50 years that torture was officially used by the state:

> Its consistent and mandatory use in the criminal justice system appears to date from 1166 and the Assize of Clarendon, according to which all those said by a jury of presentment (an accusing jury) to be "accused or notoriously suspect" of being robbers or murderers or thieves, or of receiving such wrongdoers, were to be put to the ordeal of water.
>
> (Kerr et al., 1992: 573)

In contrast, Bartlett (1988) provides quite solid ground for believing that it had been in use since 500AD and that it had certainly spread around Europe by 800AD. The forms of torture used in the trials were diverse, though the most common practices generally involved unbearable heat or drowning in water. For example, the accused might be required to grasp a hot iron for a given period, carry a hot object for a certain distance; be bound and dunked in cold water or thrown into a river. Bartlett (1988: 2–12) reports that the first trial to be put to widespread use involved a combination of the two, namely the ordeal of the cauldron, in which a large vessel was used to boil water from out of which suspects would have to pluck an object. The role of the clergyman was to determine, by reference to the burns and scars and how well they healed, whether God had intervened to protect or repair the skin of the suspect. Such intervention would indicate that God had deemed this person innocent. These various techniques of linking divine truth with the physical wound appear to have been quite standardised, and attempts were certainly made to prevent the accused persons from manipulating the process to their own ends (Pilarczyk, 1996). In this regard, trial by ordeal was not a haphazard or crazed procedure, it was not fervour or madness that framed these trials, but rather a formalised set of actions through which a given set of outcomes could be achieved and justified (Colman, 1974).

Moreover, whilst trial by ordeal certainly was used to determine culpability, it was not the default method. Historians have worked hard to explain exactly how the ordeals fit into the larger set of legal and criminal practices of the period that were more commonly used to decide legal conflicts. Bartlett (1988: 13–23), for example, argues ordeals were used to resolve disputes in which the more common procedure of swearing oaths and giving testimony were insufficient. The selection of examples that he cites also point to some of the social concerns of the period. For example, the ordeals were used in cases of adultery, to settle questions of paternity and to test religious conviction. Crucially, he argues, the ordeal was generally used only after other forms of proof had failed to provide an adequate resolution of the issues:

> The apparent diversity of the situations in which the ordeal was employed should not hide the fact that, beyond this variety, there was common ground. Sexual issues, such as adultery or disputed paternity, are, by their very nature, the cases least usually resolved by witnesses, there being no visible evidence on which to base a judgement. . . . Heresy and other cases in which trials of faith were used have a similar "invisible" quality – what was at issue was belief, and belief is intangible.
>
> (Bartlett, 1988: 33)

On this evidence, the ordeals seem to have functioned within the broader legal system, not as a displacement of justice or an alternative to it, but as an integral component of a society attempting to deal with the problem of hidden knowledge. The use of witnesses and other forms of evidence could not always satisfactorily resolve such disputes, and so it might be that the ordeal sufficed to provide an answer and a defensible outcome, nonetheless through deplorably violent means, in recognisable ways. Kerr et al. (1992) argue that the ordeals may in fact have served as a means of avoiding overzealous punishments, especially when used in cases for which the sentence for conviction would have otherwise been death. What this scholarship points towards is the idea that torture was situated within wider sets of practices that contextualise its meaning and function within the norms of the time. It was not ignorance so much as it was a compromise for those that implemented it, evidencing the limitations of the period's epistemic practices and the problems of social order that prevailed. Although cruel and abhorrent, torture was not irrational savagery.

This appears more clearly in descriptions of how the accused persons were treated during their confinement and trial. They often had to live with the clergymen, effectively participating in clerical life for three days prior to and for the three days following the physical trial (Pilarczyk, 1996). Partaking in confession, group prayers and conversations during this week meant that the priests had ample time to form an opinion on the individual's culpability before their assessment of the wounds and their official determination of the subject's guilt. Indeed, some historians have pointed out that the procedures of the trials meant that those administering them could potentially – and perhaps frequently did – manipulate the results, for example by allowing the iron to cool, by cooling the subject's hands in water before burning them, or just through their individual judgement of the wounds on presentation. As Colman (1974) argues of medieval judicial processes more generally, the trial by ordeal was a distinctly social process. It was not a simple deferral to God but mixed human and divine agency. In this way, torture can be viewed as a social technique of lie detection, responding to a problem of truth.

In whatever ways the trials fit into the practices of the period, the edict of the Fourth Lateran Council brought them to an end and eliminated state-sanctioned torture in England, at least officially. Torture remained for purposes other than determining the truth, for example in the simple form of punishment. Whilst this also seems to have eliminated trial by ordeal in the rest of Europe, other practices of torture remained a routine and visible part of the judicial systems on the Continent (Langbein, 1977). The primary use of torture in Europe was not to decide the case in the absence of evidence but rather for the development of evidence, not only in the form of confession of guilt but also corroboration. As Langbein (1977: 5) describes, 'torture was supposed to be employed in such a way that the accused would also confess to details of the crime – information which . . . "no innocent person can know."' Torture was being used as a kind of guilty knowledge test, to extract information as part of an investigation, which could be used to build a case but also to identify co-conspirators, other criminals and so forth. In this way, it served the developing mechanisms of justice without symbolic recourse to divine intervention.

Although investigative torture survived more publicly on the mainland, it took hundreds of years for it to be officially brought back to England, during the reign of Elizabeth I. In the meantime, the English legal system had been slowly reformed and the jury had begun to take on a more prominent role in determining the truth of disputed facts. Since the jury had taken up the role of deciding cases in which other forms of proof were difficult to obtain, torture could not easily be brought to bear on the same matters. The English thus borrowed from the Continent, and torture returned under Elizabeth's purview not as adjudication but as coercion, and it served historically specific purposes. At the same time, the nature of truth was also being reconfigured in the natural sciences. As Elizabeth ascended to the throne, Francis Bacon and others inaugurated the experimental method. All of which was entangled with the central spiritual question of the era: did Protestantism or Catholicism represent the true Christian faith? Discovering the truth of nature, of law and of the soul were therefore intertwined pursuits, which would gradually become part of a more general structure of social relations, as Foucault (1995) so nimbly dissects.

Elizabeth officially reintroduced physical coercion because she needed to root out Catholic traitors and plotters against her throne. In this way, torture was a response to the Protestant claim that Catholicism represented both an external and an internal threat to the nation, one which had to be eradicated. As an investigatory tool, torture in England thereby came to function as a socio-technical interrogation mechanism, rather than being an ordeal to be survived before God. But having been targeted so emphatically at a specific religious group, it was hard not to see it also as a form of punishment for one's beliefs. The lines between torture as investigation, proof and punishment were very much blurred.

In 1570, faithful Catholics were instructed by Pope Pius V, via a 'papal bull', to no longer obey the orders, mandates and laws of Queen Elizabeth, who was simultaneously decreed a heretic for her Protestantism and thereby excommunicated (Wagner and Schmid, 2011). The decree tacitly sanctioned the plots and assassination attempts that were subsequently carried out by various contesters and enemies of her rule. It was also one of several catalysts for the creation of punitive anti-Catholic laws passed by Parliament in the years that followed. Catholics became stigmatised and suspect, making them subject to a range of more or less violent practices of investigation and interrogation (Hanson, 1991). The introduction of torture as part of the demonisation of the Catholic church had the effect, ironically, of encouraging greater degrees of lying and deception amongst the faithful: the practice of 'church papism' became common for Catholics, many of whom would maintain a Protestant façade in public whilst practising Catholicism in the home (Walsham, 1999). In this way, lie detection methods did not so much ease the political weight of civilian secrets but pushed them further into the corners and recesses of social life, over which everyday people maintained some control.

As with justifications made for the torture of suspected terrorists today, religious, social and political conflicts were therefore central to bringing back into public view and reconfiguring methods of state-sanctioned torture in medieval England. Contemporary manifestations of torture organised around a frightening 'Other', which take the form of the archetypal and racist image of the Muslim as terrorist, are reflected powerfully in the form of the Catholic plotter in medieval England. National identity was just as central then as it is today to the justification of torture as an investigatory tool, for Protestantism's Other was also constructed primarily through a series of singular events:

> The failures of the Northern Rebellion, the Spanish Armada and Gunpowder Plot were absorbed into a Providential narrative of deliverance in which God periodically saved an elect Protestant nation from the assaults of the forces of the Antichrist, the last transformed into a national holiday of thanksgiving, Gunpowder Treason Day or Guy Fawkes Day.
>
> (Marotti, 1997: 38)

It was especially the threat at home which animated Protestant officials, such that an internal threat was then internalised one step further: the body became the key surface upon which a battle for the truth of Elizabethan subjects' intentions was waged.

Towards the end of Elizabeth's reign, the famous case of the Catholic priest John Gerard exemplifies the way the Privy Council used torture to compel suspected plotters to provide information. In Gerard's case the Council wished to discern the whereabouts of Henry Garnet, a compatriot, who later became infamous for his role in the Gunpowder Plot. The Council instructed the Lieutenant of the Tower of London and other legal officers:

> to examine [Gerard] strictly upon such interrogatories as shall be fit to be ministered unto him and he ought to answer to manifest the truth in that behalf and other things that may concern her Majesty and the State, wherein if you shall find him obstinate, undutiful, or unwilling to declare and reveal the truth as he ought to do by his duty and allegiance, you shall by virtue hereof cause him to be put to the manacles and such other torture as is used in that place, that he may be forced to utter directly and truly his uttermost knowledge in all these things that may any way concern her Majesty and the State and are meet to be known.
>
> (Privy Council, 1597: 38)

Contemporaneous with Elizabeth's resort to torturing the truth from the body was the emergence of experimental science, and there were ways in which each informed the other. Francis Bacon, for example, frequently invoked the use of metaphorical handcuffs and chains to uncover nature's secrets (Merchant, 2008). Although Bacon himself was careful not to use the word 'torture' to describe his scientific investigations (Pesic, 1997), some of his admirers were less cautious, Leibniz included: 'Here belongs also the art of inquiry into nature itself and of putting it on the rack [. . .] the art of experiment which Lord Bacon began so ably' (Leibniz, quoted in Pesic, 1997: 195). Of course, such thinkers' aesthetic tendencies in reporting their investigations hardly merit accusations of barbarism, but neither are the metaphorical equivalences coincidental or merely superficial. For example, Bacon's role as Special Examiner for the Crown meant that he himself oversaw the use of torture in interrogations (Black, 1927). Such complicity warrants far more condemnation than does any metaphorical flourish. In addition, the use of imagery from torture in experimentalism points to a general borrowing and back-and-forth between these seemingly disparate epistemic regimes. For example, medical science also conceived of the body as a recalcitrant but tameable source of the truth. Practitioners sought to overcome its eerie silence to save it from destruction:

> It was not that the body had to be saved from forces situated outside itself, but, rather, it had to be purged from its own internal and treasonous desire to destroy itself. That desire could be discerned in the body's refusal to speak – in its determination to preserve its tendency toward destruction through a veil of secrecy and silence. . . . The body had become the rebellious subject, hoarding its own knowledge of illness. . . . The language of treason, rebellion, conspiracy, and physical torture – a torture deployed not by the disease but by the physicians – was now determining the progress of illness. In order for the body to be saved, it had to be made to speak and confess.
>
> (Sawday, 2013: 33–34)

Alongside this symmetry with torture, Bacon, his contemporaries and inheritors also borrowed from the shifting procedures of legal evidence and the developing practices of jury trial. They highlighted the function of witnesses and the careful cross-examination of information as being similarly vital to experimental observation, inculcating social practices into the procedures of objectivity (Shapin and Schaffer, 1989). That advocates of experimentalism looked to the mechanisms of torture and to juridical procedure for instructive parallels to develop their methods reveals a broader co-production of law, science and torture, putting speech and confession at its heart.

Further support for such a conclusion can be seen in the ways in which practices of torture, now thoroughly revived and public, borrowed the materiality at the heart of variously burgeoning scientific processes (Sawday, 2013). Torture's new advocates largely assumed that the subject's silence could be overcome via actions on the human body because so many other secrets of life and illness were being extracted from its viscera (Hanson, 1991). Whilst trial by ordeal had been used primarily in the absence of proof and to elicit divine speech, torture in the Elizabethan period was used to extract concealed human truths. Rather than responding to a distrust of sworn testimony then, torture *constructed* a form of testimony. In effect this became the first scientific lie detection mechanism used for interrogation. What was invisible in the mind could be wrenched out by physical coercion: an ontological link between body, mind and pain was established. However, this came at the price of an epistemological compromise:

> In the discursive economy of English torture the body functioned amphibiously, giving truth a basis in material reality that made it susceptible to discovery, while in the intense subjectivity of its pain, making truth inaccessible to all but the sufferer.
>
> (Hanson, 1991: 59)

Having located truth in the body of the subject, torture was itself condemned to rely on confession. By encouraging suspects to speak so as to end their suffering, torture undermined its own validity. Confessions were suspect precisely because at some point persons would say anything, and likely whatever it was they thought that their interrogators wanted to hear. Catholics and a number of political commentators thus contested such use of torture (Hanson, 1991; Langbein, 1977), not only on ethical but also on epistemic grounds. They argued that the rack served to verify desperate concoctions rather than expose a hidden truth, extracting confessions that were themselves fraudulent. For Catholics, what seemed to be under investigation was less the facts and more their faith.

On this conjunction of confession, truth and torture, Bacon's views are further elucidating. He felt that torture could be justly used in the highest cases of suspected treason, but only to discover or to forestall plots and plotters, and never as a resolution of a legal dispute (Pesic, 1999). His developing approach to experimentalism thereby seems to have curtailed, at least for his investigations, the uses of statements during criminal trial which had been extracted via the rack. It also suggests that the jury had taken on more of the adjudicative role in justice and that torture was no longer to have a place in criminal trials. This would be little comfort to those subjected to it, but it is pertinent to our understanding of how knowledge and justice were being further entwined in relation to the body. In a way, this also represented the first scientific dismissal of lie detection evidence for criminal adjudication because of epistemological uncertainties. As Bartlett (1988) argues, justifications for the end of trial by ordeal had been primarily religious rather than scientific. In this later period, it was a problem with the epistemological apparatus which resulted in an at least partial dismissal of torture from the trial: you simply could not be certain that a confession was truth.

As scientific norms accrued greater powers over knowledge production, not only of nature but of life more generally, the alliance of torture and truth was thrown into doubt. But most especially it was peace between Spain and England, following the coronation of James I, that made the torture of suspected plotters generally unnecessary (Langbein, 1977). In addition, various shifts in the laws of proof following the abolition of trial by ordeal had rendered the jury trial more capable of resolving criminal disputes. Torture gradually fell out of use within trial elsewhere in Europe during the 1700s, following similar shifts in criminal procedure and campaigns by public intellectuals. State-sanctioned torture diminished[1] in England and abroad, as the techno-political landscape slowly shifted and the law found new ways not only to help govern the population but to produce facts.

The emergence of the jury trial

As torture fell in and out of official use in England, the jury system was slowly taking shape. Juries have long played a role in English criminal procedure, though not always in the way that we understand it today. As Forsyth (1852) describes in some detail, in the centuries prior to 1215 there were a variety of ways of forming a jury, and they were afforded different responsibilities over time related to different kinds of crimes and in tandem with changes occurring in the practices of trial by ordeal and the swearing of oaths (compurgation). Often the jury acted more as a 'jury of presentment': made up of men of the county, its purpose was to present to the justices of the county anyone who had been accused of criminal offences and to help to determine how the accused should be judged (Anand, 2005; Groot, 1982; Hurnard, 1941). As Forsyth (1852: 207) describes:

> Here it seems that the jury were acting rather as accusers than as triers, and at all events we see that they did not give their verdict upon evidence taken in court, but upon the private knowledge or belief which each had beforehand of the commission of the offence in question.

This is not the trial as we know it today. Rather, the most immediate issue to be decided by such a jury was generally whether the accused should face trial by ordeal or whether compurgation would suffice. If the latter, they would have to swear an oath that they had not committed the offences and produce a certain number of 'oath-helpers' who would swear to the same facts (Anand, 2005; Groot, 1982; Hostettler, 2004; Hurnard, 1941; Forsyth, 1852). A sworn oath had the effect of making those willing to lie about their testimony perjurers not only before the law but also before God, with obvious implications for their immortal souls. Colman (1974: 586) reminds us: 'Faith in divine justice was a lively reality in medieval communities as public attitudes and private actions constantly attest.' As such, if the accused persons could make this oath and rally the requisite number of oath-helpers, then it was possible that they could be absolved of the crime. The decision regarding whether the accused could proceed through compurgation or would be required to endure the ordeal depended not only on the availability of oath-helpers but also on certain social norms, as Colman (1974: 586) further articulates:

> foreigners were precluded from clearing themselves by way of wager of law, as were those individuals who had previously been suspected of criminal acts. Unfree accused could also not purge themselves by way of compurgation unless they were vouched for by their Lords. The more serious the offence, the less likely it would be for an accused to be allowed proof through compurgation.

A distrust of the previously accused, foreign and enslaved was thereby built into the practices of justice. Jurors could sometimes nullify the trial if they felt that there was not sufficient reason to believe the accused was guilty or if they distrusted the oaths sworn by the prosecution. In this regard, the criminal procedures of the time used the pathways through nullification, compurgation and trial by ordeal to mete out justice under divine decree whilst retaining significant social control over the outcomes. As with the clergy's control over the ordeal, historians generally seem to agree that the jury of presentment, in whatever forms it took exactly, seems to have had significant freedom to select the routes that suspects took through the justice system (Anand, 2005; Fisher, 1997; Groot, 1982; Kerr et al., 1992; Bartlett, 1988).

After 1215, the role of the jury began to change in England, directly in relation to the end of trial by ordeal. One significant development in the period is that of the double jury system, so that accusations were separated from hearings. The 'grand jury' heard on whether a person should be put to trial, and the 'petit jury' determined the verdict (Forsyth, 1852). For some time, however, individuals might have sat on both juries, until legislation was passed in 1352 allowing the accused to challenge the presence of a juror on the trial jury who had also sat on the accusing jury (Anand, 2005).

Whilst its configuration was changing and uncertain, the early trial jury, as Bartlett (1988) suggests, was still likely to have a mix of two responsibilities, to provide testimony themselves as well as to hear and make some judgement on evidence from other witnesses. Contrary to modern jury systems, jurors in the period were often already informed about the criminal matters at hand, since the accusing jury was required to bring the accused to the justices on behalf of the community, and jurors may well have been witnesses to the events in question. As such, the juror's roles prior to the abolition of ordeal, as witness and bearer of evidence, continued. Oaths also remained crucial, for they helped fill the gap the ordeal had left in these nascent jury-focused practices. Oaths were a way in which divine sanction could be retained within criminal procedure without the direct involvement of the clergy. By swearing an oath to God, the testimony could still be treated as a truthful account, and in this form compurgation lived on.

However, this caused a potential problem, since juries might encounter competing accounts in sworn testimony. The way around this was to forbid accused persons from swearing an oath and testifying before God on their own behalf. In fact, they were not allowed to testify at all to begin with, and it took hundreds of years before they were allowed to call their own sworn witnesses (Fisher, 1997). This early jury trial thereby ensured that it was humans that determined the fate of the accused, allowing for various possible outcomes to be subtly brought about through apparently divine procedure, avoiding the possibility of having to officially decide between competing accounts sworn before, and thus endorsed by, God. This protected faith and social order, whilst preserving a notion of truth that could sustain the justice system.

In this shifting legal terrain, there remained powerful resonances between trial by ordeal and the emerging jury trial, as Fisher (1997: 601) describes:

> The institutional brilliance of the ordeal was that it so neatly merged the appearance of divine judgment with the reality of a great measure of human control. Its verdicts at once satisfied the public desire that judgments of death be divinely ordained and flattered human notions of substantial justice. In the trial jury, the English justice system managed to reproduce this very useful combination of traits. By forbidding conflicts of oaths in capital cases,

> the system assured that a jury's verdict of guilt would seem to bear a divine imprimatur. Yet, by granting the jury an unbounded discretion to acquit, the system also assured that the trial process would not devolve into a mechanical totting of oaths, subject to manipulation by those willing to swear to false accusation.

The development of the early trial jury began the process of imbuing jurors with the power to determine a final and binding judgement on the guilt or innocence of a defendant. In the centuries between 1215 and Queen Elizabeth's reign, the English trial jury gradually became less self-informing, as it was more frequently composed of individuals who had no knowledge of the crime and its effects, and so it came to depend more upon sworn witness testimony (Langbein, 1978). In part this was because the trial jury had been separated from the accusing jury, and the accused persons had been given the right to challenge the presence of certain jurors at their trial. Moreover, there was no regular police force in the period, and so the trial jury, no longer directly informed, took on the responsibility for determining the credibility of witnesses presenting competing testimony; the accused person's own account (without a sworn oath) was compared with that of those speaking against them (often with sworn oaths) (Anand, 2005). In effect the oaths held out, but the jury was slowly becoming the trial court's lie detector, endowed with the capacity and responsibility to discern witness veracity. Truth was slowly invested in the social body, rather than the material one.

This is an argument developed in some detail by Fisher (1997), whose extensive research shows that the development of the right of the accused person to bring witnesses to speak on their behalf, even though to begin with they too were not allowed to swear oaths, meant that the jury had to determine whom to believe. The right to call sworn witnesses for the accused appears to have first occurred in England in large part due to a more practical political issue having to do with the Act to Abolish Hostilities with Scotland 1607 (Fisher, 1997: 609–615). A misunderstanding about Scotland's legal system – the English Parliament believed that Scotland allowed the accused to call witnesses – meant that in ending hostilities with Scotland, the Act, which created a merger of the English and Scottish trial systems for use in cross-border crimes, also made it possible for accused parties to call witnesses for the first time. Whatever the exact political motivations behind this first appearance of competing testimony, the Act brought with it an opening for the jury to develop a lie detection function.

However, the trial jury's determination of credibility in the face of competing testimony did not expand rapidly, but rather was curtailed and shaped using existing witness competency rules and the development of a range of new rules for guiding the jury's role within the trial (Fisher, 1997: 624–638). In effect, various rules of procedure in court appear to have been designed to protect the jury as best as possible from confrontation between two sworn parties, and where this did occur several other rules were then adopted to ensure that the truth was determined in a justifiable and – in some regards – mechanical fashion (Fisher,

1997: 638–655). This spared direct challenge of the oath as a legitimate means to ensure the veracity of witness testimony. It also shows that there was not yet enough confidence that the trial jury was competent to take on such a role and that public justice was not yet ready to let go of the faith it placed in the oath as a guarantor of true speech. It was not until accused parties were allowed to swear an oath and speak on their own behalf, a change that would not come until the late 1800s, that the jury properly began to take on the role of lie detector in England.

It is here that we make our leap across the Atlantic, since the same procedures for jury trial had by now spread and developed in the USA and it was in roughly the same period that the jury's lie detecting function developed there too, beginning in the North and gradually spreading to the South (Fisher, 1997: 662–671). Importantly, this shift in the USA was, according to Fisher, part of the Northern states' campaign to force the Southern states to abandon several racist laws, which unfairly pitted black Americans, who were deemed unable to testify at all, against white Americans, who were not only free to speak but could do so under oath. As several such laws were abandoned, defendants were simultaneously awarded the right to speak on their own behalf under oath. The jury's role as lie detector finally took hold.

As the jury slowly accrued the responsibility for judging witness credibility in the face of competing sworn oaths, it took its place at the heart of the criminal trial, becoming ever more important to the determination of the central issue at stake: were accused persons lying when they said they were not guilty? What had gradually taken shape was the criminal trial as we know it today, developing out of torture in the form of trial by ordeal, then through various manifestations of the jury of presentment, to the split of the accusing and trial juries, to the final arbiter of truth in the contemporary arrangement of criminal justice in the West. Where torture regained urgency under Elizabeth's reign in England, it had been under the guise of evidence-gathering, to forestall political plots, which was fundamentally tied to the body as the location of truth and to developing scientific principles of empiricism. The law certainly borrowed some of its evidence-gathering practices from this emerging scientific paradigm, but it was entangled with a range of political and social norms, long-established legal rules, religious structures of power and so forth. Whilst the spread of scientific norms therefore influenced the legal system's approach to epistemology, it was not the case that it disavowed other social commitments and long-standing legal tenets. Instead, the criminal trial procedure made use of experimentalism's forms of evidence and empirical reasoning as part of a broader co-production of law, knowledge, power and social order. In many respects, the jury as lie detector was not born of legal principle but of techno-politics.

Lie detection, truth and torture

The jury has not acquired this role because of a well-founded capacity to determine credibility. Instead, it found its place amidst a mix of shifting legal practices which resulted in persistent criticism of the jurors' ability to make such decisions:

> From at least the mid-sixteenth century until well into the nineteenth century, commentators often criticized the jury as being composed of individuals who were not up to the challenge and demands of the job. [. . .] Another constant criticism of jurors was that they were easily corruptible.
>
> (Anand, 2005: 422–424)

Worries and complaints about the jury's abilities and impartiality continue to this day, for example around racial composition and the effects this has on verdicts (Ellis and Diamond, 2003). Nonetheless, by the turn of the twentieth century the jury had become the heart of the Western justice system. Practices developed in England had spread to the USA, and American attorneys and prosecutors played out their cross-examinations of witnesses as a matter of routine. Experts had also become a prominent feature of the criminal jury trial in the States. Doctors and psychiatrists regularly attested to medical and mental phenomena, and a burgeoning forensic science community had also emerged (Thomas, 1999; Rafter, 2008).

Since the days of ordeal, over the centuries of development of the jury system, criminology had also been evolving. From the 1700s onwards, an array of scientific theories of crime and criminality were developed, refined, abandoned and resurfaced (Rafter, 2009). By the early 1900s proponents of forensic technologies had succeeded in their efforts to get fingerprint and handwriting analysis admitted as scientific evidence in criminal trials. These technologies and their inscriptions came regularly before the courts along with testimony from experts on the significance of their findings for the case at hand.

Each of the various forensic tools developed in the period has a complex socio-technical history. However, there is one common theme shared amongst them, in that they have all in some way promised to bring about a more scientific method of organising criminal investigation, the justice system or social life more generally. Whilst the law was reshaping the criminal jury trial, science sought to have a hand in what it became. As regards fingerprinting, for example, Cole (2002) demonstrates in detail how the technique was initially developed as a method of archiving individual uniqueness for various purposes but was transformed into one of the most commonly used procedures of criminal investigation today. From the late 1880s efforts began to turn fingerprinting into a forensic technology that would service not only the police but also the criminal trial. By 1912 it had succeeded and been made admissible, moreover the English and American courts had both established law to the effect that a person could be convicted of a crime based on fingerprint evidence alone (Cole, 2002).

Supporters of fingerprinting suggested that it 'could transform criminal justice by basing it on a stronger form of truth' (Cole, 2002: 169). So when Hugo

Münsterberg put forward his arguments for the potential of psychology to assist in the investigation of crime, he was not alone in thinking that science could be the solution to various social ills if it could be used to reconfigure legal epistemology. Münsterberg was one of several prominent figures who publicly decried the use of torture in twentieth-century policing practices. He often couched his counsel against such techniques in the context of history whilst also putting forward a promise for the future:

> The vulgar ordeals of the 'third degree' in every form belong to the Middle Ages, and much of the wrangling of attorneys about technicalities in admitting the 'evidence' appears to not a few somewhat out of date, too; the methods of experimental psychology are working in the spirit of the twentieth century. The 'third degree' may brutalize the mind and force falsified secrets to light. Enlightened juries have begun to understand how the ends of justice are frustrated by such methods.
>
> (Münsterberg, 1907: 622)

This is a rhetoric of progress which has long accompanied lie detection practices and has been central to proponents' arguments for the utility of their devices. Psychology's promise relies 'on a narrative of progress that values psychology as a forward-looking discipline capable of transforming other fields as well as society at large' (Littlefield, 2011: 74). The future, then, is as much at stake as is the present in such narratives of science's capacity to bring about justice. The future imagined by advocates of technical solutions to crime and other social problems is one in which we are governed objectively, one in which the techniques for the study of materiality are turned to the prevention and punishment of crime. In other words, each forensic claim regarding the production of truth was paired with claims regarding the production of justice.

Related to these kinds of epistemological and political propositions was an ontological one regarding the relation between the body, technology and truth, for it was generally the human body in these accounts that lent an interface between science and justice. Much like Catholics in Elizabeth's England, opponents of third degree practices of torture in the twentieth century were keen to point out that a confession wrought through violence on the body could not be trusted. What constituted the relations of truth with the body, then, was one of the central issues at stake in the advancement of lie detection and in the abolition of police torture.

How to access the truth of the body has been the central concern of lie detection experiments, which is cogently described by the existing historical examinations of the polygraph's development (Alder, 2007; Bunn, 2012; Littlefield, 2011; Weber, 2008). In the early twentieth century, several scientists, policemen and physicians variously cooperated or competed to produce a few different versions of a device that could record bodily changes and represent those changes visually. Amongst these key figures were William Moulton Marston, Leonarde

Keeler and John Larson, all of whom sought to create a machine that could tell truths from lies.

Bringing together a few different measurements, one such device eventually became known as a 'poly-graph', a machine that produced characteristic scribbles of multiple lines on paper to represent changes in bodily phenomena like blood pressure and respiration. Combining the inscriptions produced by a blood pressure cuff, pneumograph and galvanometer, for example, the polygraph drew out measurements of seemingly invisible bodily phenomena. Such devices for the measurement of these phenomena themselves have interesting histories, which cannot be fully explored here but which point to the fact that the polygraph arrived at a specific time in history not only because of political and legal forces but also due to developments in physiology, medicine and the human sciences more broadly.

Bunn (2012) weaves the invention of these early lie detecting devices into the history of psychological and anthropological theories of crime and deviance, which depended on certain claims about heritability, physiology, mental capacity and so on, whilst Littlefield (2011) and Weber (2008) show that the machines developed in relation to the scientific study of the emotions as much as they did to theories of crime. Both trajectories are important, however, since they clearly show that the pursuit of truth was tied into the ongoing study of the human body.

Citing a range of factors in the nineteenth century, including developments in statistics and public health, Bunn (2012: 8–11) describes the emergence of criminology and a concurrent shift in concepts of criminality. Whilst previously understood to be evidence of moral degeneration, criminal acts came be characterised instead by reference to naturalistic and innate factors. The connections between biology and crime were inescapably political. Charting some of this physiological concern with criminal bodies, for example, Bunn (2012: 11–20) provides an account of how the skull, in phrenology, became linked to reformist movements in the penal and judiciary systems. Phrenology's primary contribution to criminology was its 'fundamental role in creating [the] born criminal type' (Bunn, 2012: 20), thereby making ontological claims between human nature(s) and criminal behaviour fundamentally political: 'the notion of the born criminal was a rich and heterogeneous tapestry of concepts that weaved together empirical data with the wisdom of folklore and tied the utopian dream of a crime-free state to an imaginative use of scientific technique' (Bunn, 2012: 28).

Criminal anthropologists developing this account used a range of presentational forms in conveying their studies and in securing the ontological relations between criminality, security and nature. Using photographs, illustrations, graphs, tables, anecdotes, narratives and analogies, they were able to persuade their readers quite effectively (Bunn, 2012). The power of these emerging scientific accounts of the body was therefore inherently tied to the technologies' abilities to produce representations through which the phenomena could be evidenced, compared and objectified. As with many other fields, these emerging sciences of the criminal mind were dependent upon visualisation technologies acting as 'inscription devices' (Latour and Woolgar, 1986), providing material forms which could carry the truth across socio-technical locations, from the laboratory to the courtroom, for example.

Weber (2008: 68–70) illustrates how scientific work on fear was also vital to establishing a link between lying, the body and the representational capacities of the measurement devices available at the time. She argues that scientists researching the emotions had come to conceive of them as 'involuntary'. Drawing on the evolutionarily primed 'fight-or-flight' mechanism, Marston, Keeler, Larson and others were thereby able to constitute the bodily changes recorded by the polygraph as uncontrollable signs of the 'fear of exposure' of a lie. Similarly, Littlefield (2011: 50) shows: 'By standardizing the interpretive leap between physiology and emotion, mechanical records come to represent objective knowledge about the body.'

Marston's work (see Littlefield, 2011: 52–56) exemplifies this relation. Building on the research of physiologists and others, Marston conceptualised the body's responses to lying as being dependent upon an involuntary reaction to the presence of a 'deceptive consciousness'. In other words, Marston's approach required that people knew they were lying and that when they attempted to cover this up their body reacted in specific ways. The ontological claim, that the 'deceptive consciousness' produced certain responses in the body, thereby facilitated his epistemological claim, that the measurement of bodily phenomena could reveal the deceptive consciousness.

The invention of the lie detector therefore depended on an 'irony of revelation' (Schneider and Woolgar, 2012) in which the surface signs of persons' demeanour, actions and so forth were understood to mask the deeper truth in their bodies. It also depended on a certain construction of the polygraph machine as 'merely' technology, and therefore as being able to observe without intruding upon the ontological connections between lie, emotion and the body, so as to make it an objective witness to concealment, all the while itself being concealed as a primary actor in the production of that connection. This would later prove central to the lie detector's complex reception in criminal law (see Chapter 5).

In this regard, the developers of the polygraph machine conceptualised and interpreted physical responses, through technological inscriptions, as signifying underlying struggles to conceal involuntary bodily responses. The polygraph pitted the will against the body and found the will wanting. Whenever someone might be induced to fear the outcomes of their interrogation, supposed the polygraph's inventors, so might you find the body revealing those fears; without recourse to physical duress, the body still chattered away against the person's will, but in frequencies which only the machine could hear. Amplified into skittering pens on a sheet of rolling paper, the polygraph seemed to whisper the truth where previously there had only been screams or silence.

Conclusion

Truth, lying and the body have long been entangled, and torture and lie detection represent only two practices within which these knots have been tied, unravelled or further secured over time. Torture, the criminal jury trial and lie detection each represent ways of dealing with the human capacity to lie about motives, actions and intentions. Lies in trial by ordeal were a private matter which could not be wrenched out by pain. The threat of pain could, it seemed to those in the period, be used to produce a confession, as subjects ordered to the ordeal by the jury of presentment might confess to avoid physical duress. However, if they refused to confess and went to the ordeal, it was only God who officially could discern the truth from the lie, meaning that lies were not truly seen in the flesh. Rather, as Augustine argued, God witnessed the truth or falsity in the mind. He then communicated this through the human body, writing it out for the imperfect eyes of earthly representatives. Unofficially, the clergy used social practices to determine whether someone was guilty or not, meaning that the orthodox account of divine lie detection masked the reality of a social resolution to the problem of mental opacity. It would take quite some time for the jury to be invested with official warrant to determine the truth from deception, and judge the innocent from the guilty. Even today, however, sworn oaths continue to be employed in the jury trial, such that God is still invoked in relation to the limits of human perception of truth.

In tracing just some of the key points in the existing literature on the history of torture, the jury and lie detection I have also outlined the important continuities as well as discontinuities in the conceptualisation of the body at the nexus of science and law. We have seen that the rhetoric of progress inherent to many of the arguments put forward for developments in lie detection practices, broadly conceived, remain consistent. However, this should not be read as an apology for torture. Rather, it is to avoid the easy dismissal of torture as being the result of irrationality: by doing so we risk duping ourselves into the false belief that our own resort to torture in contemporary Western countries is more moral because we are more enlightened. We risk thinking that what once was the product of a society still dragging itself out of primitivism is now a regrettable necessity warranted by the evils of our time. Instead, we must see that our own use of torture and lie detection techniques are both tied to the past and specific to current socio-technical and legal problematics. It is especially important not to accept the narrative of progress with regard to lie detection, simply because it is now framed within scientific practices, for this fails to articulate the exact ways in which lie detection is positioned within the co-production of science, law and social order.

As historians have shown, lie detection emerged against a background of problems in policing, particularly the 'third degree'. Its proponents argued that technical devices could read the body's secrets without the need for duress and thereby put forward a political ontology of the body, one that – properly enacted within lie detection practices – promised to bring about a more scientific form of justice. However, the jury had been consolidated by this point as an important mechanism in the production of truth: it had become the court's lie detector.

To further understand the relations between lie detection, the law and social organisation we need to examine what happens when devices such as the polygraph machine are brought before courts. To begin to do so I now turn to the question of the status of the lie detector in the US criminal jury trial, outlining three broad positions on admissibility, before going on to address – in later chapters – the far messier picture that can be discerned from a closer look at the US state supreme courts' shifting approaches to admissibility of lie detection, and how the polygraph has become embroiled in police investigations and probationary programmes.

Note

1 Witch trials in later decades provided intense periods in which torture was brought publicly into legal spaces, which I explore in Chapter 7.

References

Alder, K. (2007) *The Lie Detectors: The History of an American Obsession*, New York: Free Press.

Anand, S. (2005) The Origins, Early History and Evolution of the English Criminal Trial Jury, *Alberta Law Review* 43 (2): 407–432.

Bartlett, R. (1988) *Trial by Fire and Water: The Medieval Judicial Ordeal*, Oxford: Oxford University Press.

Black, E.G. (1927) Torture under English Law, *University of Pennsylvania Law Review and American Law Register* 75 (4): 344–348.

Bunn, G.C. (2012) *The Truth Machine: A Social History of the Lie Detector*, Baltimore, MD: Johns Hopkins University Press.

Choudhury, S. and Slaby, J. (2011) *Critical Neuroscience: A Handbook of the Social and Cultural Contexts of Neuroscience*, Chichester: Wiley-Blackwell.

Cole, S. (2002) *Suspect Identities: A History of Fingerprinting and Criminal Identification*, Boston, MA: Harvard University Press.

Colman, R. (1974) Reason and Unreason in Early Medieval Law, *Journal of Interdisciplinary History* 4 (4): 571–591.

Ellis, L. and Diamond, S.S. (2003) Race, Diversity, and Jury Composition: Battering and Bolstering Legitimacy, *Chicago-Kent Law Review* 78 (3): 1033–1060.

Fisher, G. (1997) The Jury's Rise as Lie Detector, *Yale Law Journal* 107 (3): 575–713.

Forsyth, W. (1852) *History of Trial by Jury*, London: J. W. Parker and Son.

Foucault, M. (1995) *Discipline and Punish: The Birth of the Prison*, New York: Random House.

Groot, R.D. (1982) The Jury of Presentment before 1215, *The American Journal of Legal History* 26 (1): 1–24.

Gross, M.L. (2010) *Moral Dilemmas of Modern War: Torture, Assassination, and Blackmail in an Age of Asymmetric Conflict*, Cambridge: Cambridge University Press.

Hanson, E. (1991) Torture and Truth in Renaissance England, *Representations* 34 (Spring): 53–84.

Hostettler, J. (2004) *The Criminal Jury Old and New*, Winchester: Waterside Press.

Hurnard, N.D. (1941) The Jury of Presentment and the Assize of Clarendon, *English Historical Review* 56 (223): 374–410.

Hyland, T. (2005) Truth and Consequences, *Penn Current* (November), https://penncurrent.upenn.edu/2005-11-03/interviews/truth-and-consequences, last accessed 16/10/2017.

Jasanoff, S. (2004) The Idiom of Co-Production, pp. 1–12 in S. Jasanoff (ed.) *States of Knowledge: The Co-Production of Science and Social Order*, New York, NY: Routledge.

Kerr, M., Forsyth, R. and Michael, J. (1992) Cold Water and Hot Iron: Trial by Ordeal in England, *Journal of Interdisciplinary History* 22 (4): 573–595.

Langbein, J.H. (1977) *Torture and the Law of Proof: Europe and England in the Ancien Régime*, Chicago, IL: University of Chicago Press.

Langbein, J.H. (1978) The Criminal Trial before the Lawyers, *The University of Chicago Law Review* 45 (2): 263–316.

Latour, B. and Woolgar, S. (1986) *Laboratory Life: The Construction of Scientific Facts*, Princeton, NJ: Princeton University Press.

Littlefield, M.M. (2009) Constructing the Organ of Deceit: The Rhetoric of fMRI and Brain Fingerprinting in Post-9/11 America, *Science, Technology and Human Values* 34 (3): 365–392.

Littlefield, M.M. (2011) *The Lying Brain: Lie Detection in Science and Science Fiction*, Ann Arbor: University of Michigan Press.

Marotti, A.F. (1997) Southwell's Remains: Catholicism and Anti-Catholicism in Early Modern England, pp. 37–65 in C.C. Brown and A.F. Marotti (eds.) *Texts and Cultural Change in Early Modern England*, London: Springer.

Merchant, C. (2008) "The Violence of Impediments": Francis Bacon and the Origins of Experimentation, *Isis* 99 (4): 731–760.

Münsterberg, H. (1907) The Third Degree, *McClure's Magazine* 24: 614–622.

Pesic, P. (1997) Nature on the Rack: Leibniz's Attitude towards Judicial Torture and the "Torture" of Nature, *Studia Leibnitiana* 29 (2): 189–197.

Pesic, P. (1999) Wrestling with Proteus: Francis Bacon and the "Torture" of Nature, *Isis* 90 (1): 81–94.

Pickersgill, M. and Van Keulen, I. (2011) *Sociological Reflections on the Neurosciences*, Bingley: Emerald Group Publishing.

Pilarczyk, I.C. (1996) Between a Rock and a Hot Place: The Role of Subjectivity and Rationality in the Medieval Ordeal by Hot Iron, *Anglo-American Law Review* 25 (1): 87–112.

Privy Council (1597) *Acts of the Privy Council of England*, London: His Majesty's Stationary Office, www.british-history.ac.uk/acts-privy-council/vol27/pp26-50, last accessed 16/10/2017.

Rafter, N.H. (2008) *The Criminal Brain: Understanding Biological Theories of Crime*, New York, NY: New York University Press.

Rafter, N.H. (2009) *The Origins of Criminology: A Reader*, Abingdon: Routledge.

Sawday, J. (2013) *The Body Emblazoned: Dissection and the Human Body in Renaissance Culture*, London: Routledge.

Schneider, T. and Woolgar, S. (2012) Technologies of Ironic Revelation: Enacting Consumers in Neuromarkets, *Consumption Markets and Culture* 15 (2): 169–189.
Shapin, S. and Schaffer, S. (1989) *Leviathan and the Air-Pump: Hobbes, Boyle, and the Experimental Life*, Princeton, NJ: Princeton University Press.
Thomas, R.R. (1999) *Detective Fiction and the Rise of Forensic Science*, New York: Cambridge University Press.
Wagner, J.A. and Schmid, S.W. (2011) *Encyclopedia of Tudor England* (Vol. 1), Oxford: ABC-CLIO.
Walsham, A. (1999) *Church Papists: Catholicism, Conformity and Confessional Polemic in Early Modern England*, Woodbridge: Boydell & Brewer Ltd.
Weber, S. (2008) The Hidden Truth: A Sociological History of Lie Detection (PhD Thesis), London: London School of Economics.

Chapter 3

The polygraph machine in the United States criminal courts

> The conflicting evidence adduced by both sides is illustrative of the difficulties frequently encountered by courts in deciding questions of fact. Inventors have, from time to time, announced the perfection of devices designed to ferret out prevaricators who appear in court as witnesses. Were such an instrument available to the Court, its road would be comparatively easy. In the absence of a mechanical "lie detector", however, the Court will have to resort to its own best judgment, after observing the witnesses, their demeanor on the stand, and the plausibility of their stories.[1]

Given the court's positive disposition in *Burr & Co. v. Brigham*[2] (1936) above, developers of the polygraph could be forgiven for thinking they had good reason to be hopeful for their machine's success in the US courts. However, the law has not welcomed lie detection with open arms. But neither is the opposite the case. The polygraph machine has, in fact, long figured within United States criminal procedure. The developers of the early devices were keen for the courts to use their technologies as part of their broader aim to found criminal investigation on scientific practice. Lie detection was thus shaped with the law in mind, whether for interrogation or trial, the hope was to have an objective adjudication of veracity. Sworn oaths were of course still embedded in the 1920s/30s trial system, but the fear of damnation had waned in the centuries following the abolition of trial by ordeal and so worries over testimony continued. Since its early days in supplanting the ordeal, the jury had become the court's lie detector (Fisher, 1997). The inventors of the polygraph thus pitted their apparently objective machines against faulty human judgement as they made a first bid for its admission into a criminal trial.

They shaped their sales pitch in line with the legal politics of the period, arguing that the device was not only more accurate at detecting lies but also more humane (Alder, 2007; Littlefield, 2011; Bunn, 2012). To some degree they were successful, and the polygraph found a foothold in American society, from which it climbed to the heights of government and ultimately came to play a prominent role in global security practices. However, their efforts in the trial courts were simultaneously constrained by the success of those same rhetorical moves which had secured a market for the machine in the broader legal and political spheres.

As the polygraph experts forged ahead with their progressive rhetoric, claiming the lie detector would be more accurate than the jury, they often found the door closed to the very heart of the law. In the first hearing on lie detection expert testimony in the US criminal courts, the device was refused access, and, as is well known, the decision also set a precedent for the next seventy years on the admissibility of scientific evidence more generally.

The *Frye* ruling

The infamous trial and appeals of James Alfonse Frye in 1923[3] decided that the then named 'systolic blood pressure deception test' could not be entered as evidence in support of Frye's innocence and determined that scientific expertise should henceforth be 'generally accepted in the scientific community' before it found use in the courts. The story of the Frye ruling began a couple of years earlier than this, however, when James Brown, a 'well known colored physician' was found dead, shot in the head with a .45 calibre revolver.[4]

'Slain in his home by unknown man'[5] proclaimed the *Star*, the weekend edition of the Washington D.C. evening newspaper. Shot, presumably, by a visitor to his house, who was identified by a witness as being 'of light brown skin, twenty-four or twenty-five years old, weighing 135 pounds, and [wearing] a dark brown suit'.[6] Besides the witness description and the revolver, which had been left at the scene, investigators also recovered fingerprints from the brickwork of the wall at the front of the property. The police, family and friends offered a total of $1150 in reward for further information on Brown's murder. But there was little progress in finding the killer until nearly a year later. In late August, Detectives Jones and Jackson arrested Frye during their investigation into the forgery of a government compensation cheque. Frye confessed to the forgery as well as to the theft of a diamond ring. 'Colored, twenty-four years old',[7] he fit the witness description of Brown's killer, and – the newspapers would report – under further interrogation went on to confess to the murder of the doctor.

However, Frye retracted his confession soon after. Accounts vary as to how this came about, but the general picture seems to be that he was following the advice of his attorney. He also claimed that one of the detectives had offered to split the reward money with him if he confessed to the murder. Frye insisted that he had accepted the bribe because he had a solid alibi that would prove him innocent if worst came to worst and he was tried for murder. However, as the trial approached, his witnesses were nowhere to be found. At a loss, his attorneys decided to enlist the help of William Marston and his sphygmomanometer to attest to the falseness of Frye's confession and the veracity of his retraction. In a state with the death penalty, Frye's life hung in the balance.

On July 22 the criminal court heard on the admissibility of Marston's evidence and his expert testimony. Judge McCoy rejected both. Frye appealed the decision, but the D.C. Court of Appeals affirmed the lower court's judgement. The challenge for the court was to discern how to make legal sense of the still developing forensic science practices of experiment and technical demonstration. In making the decision, the court of appeals supplied an often-quoted epistemological determination as to why scientific expertise of this variety should be admitted or not as evidence at trial:

> Just when a scientific principle or discovery crosses the line between the experimental and demonstrable stages is difficult to define. Somewhere in this twilight zone the evidential force of the principle must be recognized, and while courts will go a long way in admitting expert testimony deduced from a well-recognized scientific principle or discovery, the thing from which the deduction is made must be sufficiently established to have gained general acceptance in the particular field in which it belongs. We think the systolic blood pressure deception test has not yet gained such standing and scientific recognition among physiological and psychological authorities as would justify the courts in admitting expert testimony deduced from the discovery, development, and experiments thus far made.[8]

The jury found Frye guilty of second degree murder, which spared his life, but sentenced him to serve the rest in prison. He was paroled eighteen years later, and by the time he was released his name had become synonymous with the tenet that scientific evidence should have 'general acceptability' before becoming admissible at trial. Although it has since been supplanted by the *Daubert v. Merrell Dow Pharmaceuticals* (1993)[9] ruling in some states, the '*Frye* standard' is still in use today and remains the principal mechanism for determining admissibility of expert evidence in some of the largest legal jurisdictions in the country.

As Alder (2007: 51) tells it, following this first failure, lie detection evidence was effectively banned from criminal trials as the *Frye* rule on expert testimony filtered across the US. It is true that for several states the typical picture was (and remains) one of rejection, but in the fine detail there is much to discern. Lie detection experts have been knocking on the door, wedging their foot in the gap or barging their way through to trial ever since. The polygraph machine has been considered for admissibility again and again in the decades after 1923 and at certain times, in certain places and for certain purposes, it has successfully gained access to the criminal jury trial in some jurisdictions. To understand this picture, it is worth first reviewing changes to evidence law which accompanied a growing dissatisfaction with *Frye*.

Expert evidence after *Frye*

During the 1960s and 1970s various efforts were made to codify evidence law to standardise procedures at the federal level. Eventually, in 1975, President Ford signed the Federal Rules of Evidence (FRE) into law, which covered evidence of all kinds, not just expert testimony. It was not until seventy years after Frye was imprisoned, however, that the United States Supreme Court reviewed the *Frye* standard and changed its position on scientific expertise. In *Daubert* the Court determined that scientific techniques entered into evidence did not have to be generally accepted in the scientific community anymore, so long as the criteria set out in the Federal Rules of Evidence were met. The case centred on Merrell Dow's drug Bendectin. It was argued by the plaintiffs – both minors (Jason Daubert and Eric Schuller) – and their parents that the drug had caused the boys' birth defects. The original trial found in favour of the pharmaceutical company, as did the Circuit Court on appeal. Both courts found that the scientific evidence submitted by Daubert and Schuller's representatives could not be admitted, since it was based on research practices that had not been in use long enough to acquire general acceptance in a relevant professional community. On hearing the final appeal, the Supreme Court held that *Frye* was now superseded by the Federal Rules of Evidence and that testimony without general acceptance in the designated community could potentially be presented to a jury.

The focus on the FRE placed far more responsibility for determining the admissibility of scientific evidence in the hands of judges. By doing so, it changed the epistemic foundations of admissibility and required judges to work as quasi-scientists (Jasanoff, 2005), for example by having to determine for themselves what counted as reliable evidence rather than deferring to the opinion of a given scientific community. *Daubert* also had the effect of 'modernising' the law, as Jasanoff (2002) argues, by introducing a greater concern for economic factors and efficiency into decision making, which had traditionally been roles for the state. The result, however, was a greater degree of uncertainty and contradiction:

> Forced largely by failures of political will in the other branches of government, courts are under pressure to see more like a state and to think more like scientists, thereby embracing two of the most powerful ordering paradigms of modernity. At the same time, judges' training in situated, context-sensitive reasoning – casuistry, if you will – keeps resurfacing at unexpected moments, so that the legal gaze is never completely transformed into the administrative gaze, and notions of individual justice are never wholly subsumed under notions of efficiency.
>
> (Jasanoff, 2002: 63)

Such tensions in the roles of judges are not always resolved in favour of state or scientific concerns. Rather, the peculiarities of a case and the justice warranted for this person and at this time may override other concerns.

As regards the polygraph machine, *Daubert* and its associated legal and socio-political impacts has had the effect of further complicating the process of determining admissibility, state by state and case by case. There is only one case in which the lie detector has been reviewed by the US Supreme Court, namely *United States v. Scheffer* (1998),[10] in which the question of how *Frye, Daubert* and the Federal Rules of Evidence should apply to the polygraph was considered. Edward Scheffer, a United States Air Force airman, voluntarily submitted a urine sample and undertook a polygraph test as an undercover informant in Air Force drug investigations. The urinalysis indicated that Scheffer had taken methamphetamine and the airman was tried by a general court-martial for multiple charges. He was convicted on all counts. During his trial Scheffer entered a motion to introduce evidence from his polygraph examination, the results of which indicated no deception when he stated that he had not knowingly taken drugs. However, Military Rule of Evidence 707[11] bars the introduction of polygraph results, polygraph examiner testimony and any reference to whether or not the accused refused or agreed to take a polygraph. As such the evidence was not introduced. Scheffer appealed but the Air Force Court of Criminal Appeals affirmed the decision. The United States Court of Appeals for the Armed Forces then reversed the lower courts' decisions. The reversal was based on the grounds that a *per se* exclusion of polygraph evidence violated the right to present a defence under the Sixth Amendment to the Constitution, part of which requires that persons accused of a crime have the right to obtain witnesses in their favour. The United States Supreme Court, as final arbiter, reversed this and affirmed the original decision stating that the *per se* exclusion did not breach the right to present a defence. It also asserted that it 'served the legitimate governmental interest in insuring that only reliable evidence was presented at a criminal trial' and did not interfere with the accused presenting any relevant fact or speaking on his own behalf.

Further comments in the Supreme Court's opinion related to the possibility that expert testimony from polygraph operators might affect how the jury made decisions on credibility. The central role of the jury, developed since the end of trial by ordeal, in determining the veracity of plaintiffs, defendants and witnesses, in effect the jury's role as lie detector of the court, was thereby put into question as experts of various kinds have sought to testify to the same issue. The result is a fear that lie detectors might usurp the jury's function, which has frequently arisen in cases considering polygraph admissibility; *Scheffer* is just one such case but one which demonstrates the way in which the law often pits the lie detector against the jury on the grounds of broader concerns in justice, rather than on the basis of accuracy, reliability or validity.

Several of the Justices also expressed the view, in accord with the Military Judge of the initial trial, that barring the polygraph served further government interests in 'avoiding litigation over issues other than the guilt or innocence of an accused,' and that restricting a defendant's right to submit evidence was reasonable in certain circumstances where other legitimate governmental interests were being served. In this regard, *Scheffer* also demonstrates the co-production of the state, justice and science in the practical resolution of criminal cases involving polygraph evidence. It shows up some of the ways in which the polygraph's exclusion from criminal trials fits into the broader context not only of the history of jury trial mechanisms, but of the production and maintenance of social order more widely, since part of what was at stake in *Scheffer* was how to govern military personnel and how to enforce drug laws. These few cases compose the backbone of national-level legal decision making on polygraph evidence. These broad decisions open up a question regarding the status of the polygraph in lower courts of the USA justice system.

Beneath the Supreme Court there are various layers of trial courts that serve federal districts, states and then regions within the individual states. There are complex relations between each layer, and differences across the country are difficult to pin down in any hierarchically consistent fashion. Slicing through the middle by looking state-by-state makes for a detailed enough picture without becoming bogged down in the vast number of cases heard in courts at the very lowest levels of criminal trials. The decisions of the state supreme courts make for a useful set of documents in which to understand how the cases reviewed so far in this chapter, those decided at the highest levels of the law, filtered down into the lower courts and affected decisions on the polygraph's admissibility, whilst also allowing for some of the specificity of state cases to shine through. This will help evidence how the status of the polygraph has differed over time and across jurisdictions based significantly on different responses to *Frye* and other US Supreme Court rulings, but also on the development of federal and state rules of evidence, along with precedents responding to particular issues arising in the many thousands of cases in which testimony has been heard regarding the polygraph's admissibility.

The polygraph machine in the state supreme courts

A heuristic is sometimes put forward that in the US there are now three general positions taken by the state supreme courts on whether the polygraph is admissible or not: (1) a *per se* ban, generally following *Frye*, makes it inadmissible; (2) it is admissible but only with 'prior-stipulation'; and (3) it is admissible without prior-stipulation. I will, in the breakdown of the chapter structure, adopt this rough heuristic. However, I also want to show that the history of cases, appeals and state supreme court decisions state-by-state evidences how the polygraph continues to probe at the edges of the trial, wriggles through the gaps and takes root in ways that do not neatly represent these broad positions. As I take each of the three sections, I nuance the position, showing that the way in which lie detection evidence is dealt with is complex and oftentimes remains uncertain.

1. Per se *ban*

States that today follow *Frye* more generally on scientific evidence do tend to enforce a blanket ban on polygraphy and recently have come to reference *Scheffer* as a modern counterpart. This is evident, for example, in Illinois, where the ban on admissibility has long been enforced using *Frye* and now *Scheffer*, e.g. in *People v. Jefferson* (1998).[12] Other states in which *Frye* has been superseded tend to follow the FRE and *Daubert* to implement their ban. This is so in Connecticut, for example. In *State v. Porter* (1997),[13] where the then recently decided *Daubert* ruling was used to supplant *Frye* but to continue the ban on polygraph admissibility. *Frye*, *Scheffer*, the FRE and *Daubert* are thus all currently being used together or separately in some US state supreme courts to uphold bans on the polygraph within criminal trials.

Where the *per se* ban remains in place and appears to be well enforced through these rulings, there are nonetheless regular appeals made against decisions to bar access by the lower courts. Such appeals are, for the most part, easily closed down in the superior courts by referring to existing precedents. The following example from Colorado is representative of rejection without having to belabour the issue:

> Dunlap next contends that he was unconstitutionally deprived of the opportunity to submit polygraph evidence in order to impeach the testimony of a prosecution witness. . . . Such evidence is per se inadmissible under Colorado evidentiary law, we hold that the trial court did not commit a constitutional error or abuse its discretion by refusing to allow Dunlap to introduce the evidence.[14]

As *Colorado v Dunlap* (1999) makes clear, some states implement a general ban with ease. However, even in these states the polygraph might still feature in cases as direct evidence not of deception or guilt, but of some other issue. Polygraph tests often find their way into witness testimony in the absence of any specific request to hear testimony from a polygraph expert. Courts then must determine whether, during the trial, such references to polygraph tests and their outcomes should constitute a breach of the general ban against their admissibility as scientific evidence.

For example, in California, a court recently had to determine whether it was acceptable for the prosecution to enter evidence regarding a witness 'failing a polygraph test', not in respect of the central issue of guilt but as context for the witness's state of mind during his interview with the police. A difficulty that appeal courts face in this regard is to determine exactly how significant such an error is if one has occurred. Judgements on this issue appear to be determined by reference to the broader context of the case. The witness in *The People v. Crandell McKinnon* (2011)[15] claimed that he had no information about two murders under police investigation. He continued to deny any knowledge of the crimes during an interview in which he was placed under significant pressure to take a test. He agreed and after being informed that he had 'failed the polygraph' he admitted that he had seen the murders and implicated Crandell McKinnon as the culprit. The California Supreme Court found that an error had occurred in allowing the prosecution to introduce the evidence that the witness had failed a polygraph test just before making his admission, even if this evidence did not bear directly on McKinnon's guilt. However, the court found that there was 'no reversible prejudice'. In other words, the error was not significant enough to warrant any reversal of the lower court's decision that McKinnon was guilty of two counts of first degree murder, because – given the broader context of the investigation – the polygraph evidence could not have substantively changed the jury's interpretation of the witness's statements or their judgements of his credibility, and nor could they reasonably be said to have led to a different outcome for McKinnon.

New York state, which continues to enforce a *per se* ban on admissibility, provides similar examples. In a recent case it was held that:

> even if the admission of this [polygraph] evidence was [in] error, the error was harmless, as there was overwhelming evidence of the defendant's guilt and no significant probability that any error in this regard contributed to his conviction.[16]

These cases show that admission of evidence of polygraph examinations and their results might sometimes be admitted even in states with bans on polygraph tests, so long as their admission is not prejudicial to the central issue.

More problematically, states must also deal with what I call 'polygraph talk', not simply admissions of evidence. This occurs through the testimony of witnesses (sometimes the defendant) who have been subject to the polygraph, or of police officers who have used it in the conduct of their investigations. Whilst polygraph experts might not be able to testify before the courts as to the polygraph's functions, validity or reliability in such cases, and whilst the jury is not exposed to material evidence of polygraph tests or results, the use of polygraphs within police investigations means that they might still play a role in a trial.

An early case in which this problem is well exemplified is to be found in *People v. Parrella* (1958), heard before the California Supreme Court.[17] During the direct examination of the appellant by his counsel (Mr. Racanelli), the following exchange occurred:

MR RACANELLI: Did you remain in that cell all that day and night?
MR PARRELLA: Up until that night. I volunteered for a lie test, I wanted to take a lie test, wanted to clear it. That was the only time I was out of the cell that day.
MR RACANELLI: Did you take that test?
MR PARRELLA: Right.
MR RACANELLI: On that day?
MR PARRELLA: Right.[18]

The prosecutor, Mr. Schatz, made no objection to this testimony. Instead, on cross-examination Schatz attempted to elicit the results of the test which the defendant had admitted to having taken:

MR SCHATZ: All right. And while you were at the police station, you told them that you weren't guilty of this heinous offense, isn't that correct?
MR PARRELLA: I did.
. . .

MR SCHATZ: Do you know the results of the lie detector test?
MR PARRELLA: No.
[At this point Racanelli, counsel for Parella, objected:]
MR RACANELLI: Counsel knows better. I don't care what the results are, for or against, it's not allowed in any Municipal Court of law. I ask he be cited for misconduct.[19]

Since the defendant himself had brought the polygraph into testimony, the judge was initially tempted to admit the polygraph talk:

THE COURT: The witness on the stand testified on direct [examination] that he had offered to take a lie detector test.
MR. SCHATZ: He did.
THE COURT: Then on cross-examination has he not a right to follow that up and ask if he did and what the results of it [were] or anything of that sort?[20]

Racanelli, however, argued that even the mention of the polygraph led inexorably to conclusions about the results and, by inference, about the defendant's guilt:

MR. RACANELLI: I submit, Your Honor, as Your Honor knows, the law is clear in this state and I don't know what the lie detector is, for or against, to be quite frank, Your Honor, but it's inadmissible in any event and this may delude the jury into thinking it should be produced and there's some reason why it isn't, all to the detriment of the defendant.
THE COURT: Of course, a lie detector test is not admissible in evidence.
MR. SCHATZ: That is true. I would never have brought it up if it wasn't offered in his direct testimony. The man said he wanted a lie detector test.
MR. RACANELLI: He can ask any question about that, he cannot ask any question about what the results are.
MR. SCHATZ: I will withdraw the question.
MR. RACANELLI: Once it's inserted, the damage is done. I ask the jury be asked to disregard the comments of Counsel.[21]

Racanelli's claim that the 'damage is done' merely by mention of the polygraph machine and its results emphasises the way in which juries might be tempted to conclude something regarding guilt or innocence on the basis of lay understandings of lie detection. In this regard, the success of the polygraph machine outside of courtrooms in the United States, a product of its representation as an infallible device in films, television shows and novels (Littlefield, 2011), posed and continues to pose a problem for trials in which it has played a role but may not be spoken of, i.e. those in which a *per se* ban is applied. Courts feared that the fictions which circulated in American culture, constructing an image of the polygraph as an objective, reliable and legally admissible technique, might influence jury deliberations simply by its mention. In this regard, the polygraph's success on screen and in the pages of magazines was also, in some states, toxic to its recurring legal battles to regain some credibility within criminal trials.

Courts negotiate such potential errors in testimony by reference to the totality of the evidence against the defendant and context of the error. They must subtly determine how testimony about polygraph tests might influence jurors' perceptions of the facts and of the veracity or credibility of witnesses with reference to the sum of submitted evidence, cross-examinations, expert testimony and so forth. The threat of such polygraph talk is taken very seriously. An en banc reconsideration of an appeal of conviction for assault with intent to murder heard in Florida[22] concluded that the court had committed a damaging error in allowing the prosecution to bring reference to the polygraph into the trial. The victim, Holton Newbold, in answering the prosecution's questions, testified that he had taken a lie detector test during the police investigation of his assault to confirm his account. The Florida Supreme Court concluded:

> When the trial judge allowed the evidence to be admitted, and sanctioned the submission by the statement he made, we are of the view that he committed as egregious an error as if he had admitted the *results* of the test. . . . For there can be no doubt that in initiating the inquiry the prosecutor intended to leave in the minds of the jurors the impression that because the witness Newbold had voluntarily submitted to a lie detector test prior to the time of trial he was a man of veracity and hence was telling the truth from the witness stand, no matter how inconsistent his tale might appear to be to the jurors when compared with the testimony offered by other State witnesses.[23]

In this case and in others, polygraph talk is treated as though the machine itself had been admitted, for it threatens to lead the jury towards conclusions through construction of certain witnesses as more or less trustworthy depending upon whether or not they volunteered to take polygraph tests during criminal investigations. For the Florida court, this results, 'in effect, in the substitution of a mechanical device, without fair opportunity for cross-examination, for the time-tested, time-tried, and time-honored discretion of the judgment of a jury as to matters of credibility'.[24] Even reference to the polygraph is therefore banned in some cases and can result in new trials, as was the outcome of *Kaminski*. The most frequent outcome of an objection to polygraph talk, however, is jury admonishment:

THE COURT: Ladies and gentlemen of the jury, the question of lie detector has no place in the case. It has been determined that lie detectors are not admissible in evidence in a trial of a case. Now, the question that the District Attorney has been asking this witness up to date was proper because the witness on direct examination stated that he offered to take a lie detector test and the District Attorney then had a right to ask him whether he had taken it or not.

> But beyond that, we can't go. So just confine your consideration of the matter to those two questions.[25]

Appeals and frequent jury admonishments because of polygraph talk have led to *per se* ban courts becoming sensitive to situations in which the polygraph might be brought into cross-examination. Having spent fifty years dealing with this issue, the California court suddenly found itself engaged in a farcical attempt to persist through cross-examination whilst also erasing the polygraph from the witness' account.[26] Things seemed to be going well to begin with, for having realised that the polygraph might be mentioned if the prosecution continued a line of questioning, the attorneys conferenced with the court to prevent the problem arising. After some discussion at the sidebar, the parties agreed on a 'sanitised' phrase, one that would substitute for 'polygraph examination': they would call it 'an appointment in Sacramento'.[27] However, this confused the witness, and the prosecution found it difficult to ask questions which did not prompt talk about lie detection, inevitably – trying to work out what he was being asked – the witness blurted out, 'You mean the polygraph?' and the jury had to be admonished to remove from their minds the idea that the polygraph had played any role in this case.

These examples of polygraph talk show that the *per se* ban implemented in certain US states is not free from uncertainties in application. Polygraph talk and admonishment suggest the ban might not be so impenetrable a barrier as it may seem, but neither is it a chalk line, easily erased. It is an occasionally flexible rule that can be bent, but only to a certain extent, and always to the demands of doing justice within a specific context, given the history of precedent in the state and the specific trouble caused in the case at hand by reference to the machine. The courts do not implement the rule about polygraphy blindly. They are spaces in which the specificity of a case must be used to guide how the rule should be implemented. But it is also this commitment to the case at hand which allows for the polygraph to continue knocking on the court's door, ceaselessly requesting re-evaluation of its admissibility.

2. Prior-stipulation

The second general position adopted in many US states is that the polygraph is 'admissible with prior-stipulation'. It refers to a situation in which both parties to the criminal investigation, the suspect and the police, or the defendant and the prosecution, agree, generally in advance of the administration of the test, that the results will be admissible as evidence at any subsequent trial, irrespective of whether the interpretation of the polygraph charts points to guilt or innocence. It is, in this way, something of a gamble for both parties. This is an often-overlooked feature of the polygraph's legal portrait and is much more than a fine line in the broader picture of its legal status. Rather, it plays an increasingly prominent role in how admissibility of lie detection is managed in US courts.

One of the earliest cases in which prior-stipulation formed part of a case at the state supreme court level was *LeFevre v. State of Wisconsin* (1943).[28] In January 1941, twenty years after the *Frye* ruling, Frank LeFevre was tried for murder and found guilty. He appealed the decision to the Wisconsin Supreme Court. As part of his appeal LeFevre claimed that the lower court had made an error in excluding polygraph evidence and testimony based on two tests he had voluntarily taken during the police investigation. The police arranged the first test with Joseph Howard Matthews, Professor of Chemistry and Head of Department at University of Wisconsin, known especially for his work on firearms identification but also for his experiments in lie detection and narco-analysis. Before taking the test, LeFevre was taken to meet the district attorney.

The DA was intent on securing evidence against LeFevre and, probably knowing that polygraph evidence was unlikely to be admitted at trial, cleverly formulated a declaration for LeFevre to sign to the effect that the results should be admissible whatever their outcome:

> It is further stipulated and agreed by and between the said Frank LeFevre and S. Richard Heath (District Attorney) that any fact, matter or thing disclosed by said lie detector examination of said Frank LeFevre and the findings thereon, may be admitted in any trial or preliminary examination before any of the courts of the County of Fond du Lac or State of Wisconsin.[29]

The DA was not satisfied with the results of the first test, which apparently corroborated LeFevre's account of his innocence. Suggesting that the police would have to 'look elsewhere' if he passed another test, the DA convinced LeFevre to go to Chicago for a second examination, this time to be conducted by one of Leonarde Keeler's assistants. In fact, three lengthy tests were performed and the results confirmed the results from Prof. Mathews' original exam. Later that year, after LeFevre was tried for the murder, he sought to admit the results from both tests along with expert testimony from Mathews and Keeler to support his case. The judge determined that *Frye* stood and the results of the tests could not be admitted. On appeal the court agreed that the polygraph results were inadmissible but found that the district attorney's testimony about these events, to which the prosecution raised no objection, could be considered in its decision on whether there was reasonable doubt as to LeFevre's guilt. The Wisconsin Supreme Court found that there was in fact reasonable doubt and ordered LeFevre to be discharged from custody. Prior-stipulation thus proved unsuccessful on its first hearing before the US Supreme Court but may have played a role via the DA's testimony as polygraph talk.

The earliest instance in which prior-stipulation was at issue and in which the polygraph evidence was subsequently admitted came five years later in California, in *People v. Houser* (1948).[30] This was not only the first case to consider the issue of prior-stipulation but also the first case before the California Supreme Court to consider polygraph evidence more generally. In a lower court, William Houser had been found guilty of sexually assaulting a girl, aged eight years. During the investigation, he had signed a prior-stipulation agreement similar to LeFevre's, but also including words to the effect of securing the polygraph examiner's credentials. Mr Reidel, a polygrapher employed by the Berkeley Police Department, used Keeler's device to examine Houser. The results – he would testify – showed Houser was lying about his innocence. The lower court admitted the evidence:

> The machine was displayed to the jury and its mechanism and the manner in which the test was given was explained and interpreted to the jury. He [the examiner] explained that he asked the defendant a series of 10 questions to be answered by "yes" or "no." Some questions related to matters immaterial to the charge and others were directly pertinent to it. The graph was introduced in evidence. The expert concluded and stated that he was of the opinion from the test given defendant that defendant Houser was not then telling the truth in respect to the accusations made.[31]

On appeal Houser claimed that the polygraph tests now had no evidentiary value and that it was an error for the court to admit the results. The California Supreme Court, however, found:

> It would be difficult to hold that defendant should now be permitted on this appeal to take advantage of any claim that such operator was not an expert and that as to the results of the test such evidence was inadmissible, merely because it happened to indicate that he was not telling the truth.[32]

The judgement was affirmed and the polygraph evidence sanctioned because of Houser's prior-stipulation. The court determined that it was for the jury to appraise the value and determine the weight of the lie detection evidence where prior-stipulation had been made.

In the decade that followed Houser's appeal, the California Supreme Court considered polygraph evidence several times over, slowly refining a position of *per se* inadmissibility. Nonetheless, the use of prior-stipulation rapidly gained popularity in US police procedures and slowly evolved in several other states to function as a kind of middle ground between the *per se* ban that *Frye* had established and outright admissibility. Again, the picture is murky, however, as some verdicts based on prior-stipulation have been overturned on appeal, for example in Illinois:

> such a debatable expert opinion, with such obvious prejudicial impact, ought not to be introduced by virtue of the oral stipulation presented in this case. While a defendant may understandingly stipulate to much in a criminal trial, and may waive many objections, we think it manifestly unfair to bind him by a stipulation regarding the trustworthiness of scientific opinion far beyond his expected ken.[33]

As with the *per se* ban, prior-stipulation continues to be challenged in courts on multiple grounds. Arguments presented by both the prosecution and the defense claim that lie detector evidence should not be admissible even with prior-stipulation, whilst others support its continued use, depending generally upon whether the evidence is favourable or not to their case.

3. *Admissible without prior-stipulation*

During the twentieth century polygraph evidence was admitted to various criminal courts in the US without prior-stipulation, though only on an intermittent basis. This seems to have happened during a period of more general confusion regarding scientific expert testimony in the US and as dissatisfaction with *Frye* grew between the 1960s and 1980s. Over the course of these few decades various efforts were made to codify evidence law, and the Federal Rules of Evidence were gradually thrashed out and subsequently promulgated throughout the states. How the polygraph fared during these years is a complex story and has to be understood state by state.

California and New Mexico make for an excellent comparison in this regard. In California *Frye* had rule over evidence law and was used to ban polygraph testimony from criminal courts for several decades despite the state being a hotbed of polygraph research and entrepreneurialism. As discussed above, *The People v. Houser* was the first California Supreme Court case in which the polygraph was denied admissibility despite Houser's prior-stipulation. *Frye*, by now the default ruling on scientific expert testimony, helped to keep the polygraph out of the state's criminal courts. Nonetheless, and likely given the research being conducted in the state, attempts continued to be made to gain admissibility over the following decade.[34] These State Supreme Court cases likely hide a significant number of lower cases which never reached the higher court.

Still, in California the *per se* ban remained largely in place until the 1980s. In 1967 the state had formulated its own rules of evidence, which bore significant resemblance to the developing position nationwide and would eventually come to constitute the Federal Rules of Evidence. The first state supreme court case heard after the shift in 1967, the trial of *Witherspoooon v. The Superior Court of Los Angeles County* (1982),[35] thus drew upon the state rules of evidence to displace *Frye* and brought about changes to how California dealt with polygraph expert testimony. Hearing Witherspoon's appeal against his conviction for eight counts of armed robbery, the California Court of Appeal determined that he was correct in his claim that the lower court should have held a pre-trial evidentiary hearing on the validity and admissibility of the polygraph machine. The court referred to California Evidence Codes 402–406, which now provided for the pre-trial hearing of admissibility of evidence relating to a disputed preliminary fact, and also to rules 801–805, which concerned how the courts should evaluate expert testimony. Section 805 was especially salient, as it stated: 'Testimony in the form of an opinion that is otherwise admissible is not objectionable because it embraces the ultimate issue to be decided by the trier of fact.' This language is a direct reflection of what would come to be the Federal Rules of Evidence section 704, which later played a significant role in the dissent of Justice Stevens, reviewed above in *United States v. Scheffer*.

On the basis of these rules, the appeal court argued:

> In our opinion, the more serious defect in the later cases which have, with an almost "knee jerk" response, continued some 60 years since the original *Frye* decision, to label the results of the polygraph examination as "unreliable," is that courts in more recent times have merely subjectively favoured one side of a dispute in which there is a substantial and credible body of opinion on both sides of the question.[36]

The *Witherspoon* case thereby opened the door to the polygraph, although of course admissibility was not simply a given. Admissibility of polygraph evidence would have to be heard on a case-by-case basis. A year later, Proposition 8 was passed in California, part of which included a constitutional amendment barring the exclusion of relevant evidence. Certainly, the polygraph was relevant to a great many issues, not least the central issue of whether the defendant was guilty or not. Together, *Witherspoon* and Proposition 8 caused a huge surge of non-stipulated lie detection examinations to be presented before courts in California. District attorneys urged the legislature to enact a new statute in the Evidence Code, which tightened up the ruling and moved California away from its seemingly accidental permissiveness with respect of polygraphy and towards the position being adopted by other states at the time, namely prior-stipulation.

The period of national instability in evidence law and the slow concretisation of the FRE had a significant though erratic impact on polygraph admissibility in California, and in other states as well. How these shifts in evidence law were negotiated was importantly shaped by the local contexts in which they occurred, as shown for example in California's development of Proposition 8. Prior-stipulation became the default position adopted in a number of other states as uncertainties regarding the FRE and the transition from *Frye* to *Scheffer* bubbled over. New Mexico is the only state to have emerged out of this period with a strong set of rules regarding polygraph evidence which lean in favour of admissibility without need of prior-stipulation. As in many other states, a *per se* ban had been in effect in New Mexico up until these decades of transformation. The *per se* ban was certainly still in effect in the early 1960s, as can be seen for example in *State v. Trimble* (1961).[37] However, in 1970, the case of *State v. Turner*[38] resulted unpredictably in a surfeit of success for several long-contested technologies. The court allowed not only polygraph evidence and expert testimony to be admitted but also hypnotism and truth serum (sodium amytal). All three were subject to a prior-stipulation, signed by the prosecution and Ernest Turner, the defendant. Each of the experts concluded that Turner was telling the truth when he said that he was innocent of aggravated burglary and battery. However, witnesses also testified under oath to identify Turner as their assailant. Given that the experts' testimony supported his case, on appeal Turner did not contest the admissibility of these lie detection techniques. However, he did challenge the way in which the experts had been cross-examined by the prosecution. The appeal court found against Turner on this matter, and the expert testimony as well as the cross-examination were allowed to stand. This did not begin a free-for-all on lie detection admissibility that it superficially permitted, however, since prior-stipulation had been made in Turner's case and thus informed the court's reworking of its procedures immediately following. Indeed, standards began to be formulated that would be applied to determine the basic requirements to warrant admissibility of evidence from a polygraph examiner. This meant that some experts did not meet the measure, for example in *State v. Alderete* (1974).[39] Although ruling the evidence inadmissible in this specific case, the decision concerned the discussion of exactly what standards should be required to qualify a polygraph expert to testify, and whether standards could be formulated as a blanket rule or admissibility considered strictly on a detailed, case-by-case basis depending upon the exact qualifications, experience and position of the expert. In beginning these discussions New Mexico was taking an increasingly lenient attitude to polygraphy. A concurring statement from Judge Lopez added that 'polygraph testing is potentially of great value to the judicial processes of this state. When we are presented with a proper record meeting the requirements set forth in Judge Wood's opinion, I would hold this type of evidence admissible. I encourage counsel in future cases to develop such a record.'

A couple of years later in *State v. Lucero* (1974)[40] a set of rules were formulated, based on those ventured in *Turner*, for determining that polygraph evidence was admissible:

> 1. The tests were stipulated to by both parties to the case; 2. When no objection is offered at trial; 3. When the court has evidence of the qualifications of the polygraph operator to establish his expertise; 4. Testimony to establish the reliability of the testing procedure employed as approved by the authorities in the field; and 5. The validity of the tests made on the subject.[41]

A year after *Lucero*, in *State v. Dorsey* (1975),[42] the state supreme court relaxed these rules significantly, providing an unusually short opinion for such matters. This removed the need for prior-stipulation and dropped the requirement that there should be no objection from either party. Polygraph evidence could thus be admitted without mutual stipulation and over objection at trial, so long as rules 3 to 5 were followed. The court's reasoning was that these two requirements did not fit New Mexico's rules of evidence and that they were contrary to the spirit of the rules as 'to secure fairness in administration and promotion of growth and development of the law of evidence to the end that the truth may be ascertained and proceedings justly determined.'

Since 1984 the New Mexico rules of evidence have further refined the admissibility of polygraph evidence and now provide detail on what requirements must be satisfied before judges have admissibility at their discretion. They include rules about examiners and the examination process. For instance, they require that persons voluntarily submit to examination, that they be examined three times, that at least two relevant questions are asked of them, and that the pre-test interview and the test phase be recorded by audio or video.

Conclusion

In reviewing the status of the polygraph machine in the late 1980s,[43] after the various developments had occurred around the states' rules of evidence and refinements had been made as to how testimony on the results of a polygraph test could be considered for admissibility, Alaska Supreme Court Justice Stowers remarked that in his state the lie detector had a 'twisted history'. He was quite correct that the status of the polygraph's admissibility had been through various turns in his territory, but he might equally have said that this was the case in the United States more generally. To my knowledge, nowhere else in the world has the polygraph machine had such a tangled history with the criminal jury trial, evidence law and questions of policing, and nowhere else is it so frequently brought before criminal courts.

The three broad ways in which polygraph evidence is dealt with in states in the US today are that it is either banned, admissible with prior-stipulation or admissible under certain rules of evidence. Whilst this kind of summarising of the three positions in the state supreme courts is useful, it also erases the historical differences between each state, masking the twists and turns that lie detection has taken since James Alfonse Frye was arrested for murder. This heuristic also does little to explain what it is that happens exactly when polygraph evidence comes into consideration within a jury trial: how is the polygraph machine understood and with what implications for legal concepts of truth and for the purpose of the jury in determining the truth?

The conditions producing these twists and turns in the lie detector's life within the courtroom have created a scenario in which legal and scientific arguments in support of or against the polygraph continue to be revised, refined, recognised and refused, producing a certain amount of instability both within trial and in the periphery. This is the point which I further develop in the rest of this book, to argue that uncertainties and instabilities both in lie detection technology and in legal practices are related in ways which bear on the meanings of lying and its socio-political management.

Notes

1 *C. R. Burr & Co., INC., d.b.a. vs. Laurence S. Brigham*, 3 Conn. Supp. 277, 278 (Conn. Super. Ct. 1936).

2 Ibid.
3 *Frye v. United States*, 293 F. 1013, 1014 (D.C. Cir. 1923).
4 Author Unknown (1921) Slain in His Home by Unknown Man, *The Sunday Star*, November 28th, http://chroniclingamerica.loc.gov/lccn/sn83045462/1920-11-28/ed-1/seq-1/, last accessed 16.10.2017.
5 Ibid.
6 Ibid.
7 Author Unknown (1921) Confesses to Murder of Colored Physician, *The Evening Star*, 23 August, http://chroniclingamerica.loc.gov/lccn/sn83045462/1921-08-23/ed-1/seq-5/, last accessed 16.10.2017.
8 *Frye v. United States*, 293 F. 1013, 1014 (D.C. Cir. 1923).
9 *Daubert v. Merrell Dow Pharmaceuticals, Inc.*, 509 U.S. 579 (1993).
10 *United States v. Scheffer*, 523 U.S. 303 (1998).
11 Rule 707: 'Notwithstanding any other provision of law, the results of a polygraph examination, the opinion of a polygraph examiner, or any reference to an offer to take, failure to take, or taking of a polygraph examination, shall not be admitted into evidence.' *United States v. Scheffer*, 523 U.S. 303, 306–7 (1998).
12 *The People of the State of Illinois v. Norma Jefferson*, 184 Ill. 2d 486, 705 N.E.2d 56 (1998).
13 *State of Connecticut v. Christian E. Porter*, 241 Conn. 57, 698 A.2d 739 (1997).
14 *The People of the State of Colorado v. Nathan Jerard Dunlap*, 975 P. 2d 723, 755 (1999).
15 *The People v. Crandell McKinnon*, 52 Cal. 4th 610; 259 P. 3d 1186 (2011).
16 *The People of the State of New York v. Michael Strzelecki*, 968 N.Y.S.2d 196, 108 A.D.3d 644, 645 (2013).
17 *The People v. Robert Carmen Parrella*, 158 Cal. App. 2d 140, 322 P. 2d 83 (1958).
18 Ibid., 86.
19 Ibid., 86.
20 Ibid., 86.
21 Ibid., 86.
22 *Kaminski et al. v. State*, 63 So.2d 339 (Fla. 1953).
23 Ibid., 341.
24 Ibid., 341.
25 *The People v. Robert Carmen Parrella*, 158 Cal. App. 2d 140, 322 P. 2d 83, 87 (1958).
26 *People v. John Sapp*, 31 Cal. 4th 240, 73 P. 3d 433 (2003).
27 Ibid., 475.
28 *LeFevre v. State of Wisconsin*, 242 Wis. 416 (1943).
29 Ibid., 424.
30 *People of California v. William Jack Houser*, 85 Cal. App. 2d 686, 193 P. 2d. 937 (1948).
31 Ibid., 940.
32 Ibid., 942.
33 *The People of the State of Illinois v. Robert Zazzetta*, 27 Ill. 2d 302, 189 N.E.2d 260, 264 (1963).
34 See, for example, *The People v. Wallace A. H. Wochnick*, 98 Cal. App. 2d 124, 219 P. 2d 70 (1950), *The People v. Lloyd Porter*, 99 Cal. App. 2d 506, 222 P. 2d 151 (1950), *The People v. Thomas Carter*, 48 Cal. 2d 737, 312 P. 2d 665 (1957) and *The People v. Robert Carmen Parrella*, 158 Cal. App. 2d 140, 322 P. 2d 83 (1958).
35 *Gary Witherspooon v. The Superior Court of Los Angeles County*, 133 Cal. App. 3d 24, 183 Cal. Rptr 615 (1982).

36 Ibid., 618.
37 *State of New Mexico v. J.T. Trimble*, 362 P. 2d 788 (N.M. 1961).
38 *State of New Mexico v. Ernest Turner*, 468 P. 2d 421 (N.M. 1970).
39 *State ofNew Mexico v. Gilbert Alderete*, 521 P. 2d 138 (N.M. 1974).
40 *State of New Mexico v. Tomas Lucero*, 526 P. 2d 1091 (N.M. 1974).
41 Ibid., 1093.
42 *State of New Mexico v. Sammy T. Dorsey*, 539 P. 2d 204 (N.M 1975).
43 *B&W Construction Company, Donald Paul Wood and Robert Bowers v. N.C. Ribble Company (defendant and counterclaimant-Apellant) v. B&W Construction Company, Donald Paul Wood and Robert Bowers*, 1987-NMSC-0l9, 105 (N.M. 448).

References

Alder, K. (2007) *The Lie Detectors: The History of an American Obsession*, New York: Free Press.

Bunn, G.C. (2012) *The Truth Machine: A Social History of the Lie Detector*, Baltimore, MD: Johns Hopkins University Press.

Fisher, G. (1997) The Jury's Rise as Lie Detector, *Yale Law Journal* 107 (3): 575–713.

Jasanoff, S. (2002) Science and the Statistical Victim: Modernizing Knowledge in Breast Implant Litigation, *Social Studies of Science* 32 (1): 37–69.

Jasanoff, S. (2005) Law's Knowledge: Science for Justice in Legal Settings, *American Journal of Public Health* 95 (S1): 49–58.

Littlefield, M.M. (2011) *The Lying Brain: Lie Detection in Science and Science Fiction*, Ann Arbor: University of Michigan Press.

Chapter 4

The exclusionary toolkit

> The lie detector records with graphic certainty the fluctuations of the pulse. There is no instrument yet invented that records with equal certainty the fluctuations of the mind.
>
> (*People v. Zackowitz*, 1930)[1]

If the door to the court was thought to have been firmly closed by *Frye*, as might be seen to have been confirmed in the *Zackowitz* decision, then subsequent decisions on polygraph evidence show that it has rather remained ajar, buffeted to and fro by restless legal and scientific winds. Cases at all levels and across all states evidence the ways in which the polygraph is sometimes admitted as evidence in criminal trials. As the previous chapter demonstrated, there have been significant changes over time and there are differences state by state in how such evidence is received by the courts. Having fleshed out some of the ways in which the polygraph is barred from or enters criminal jury trials, in this chapter I develop a deeper appreciation of the complexities of how polygraph evidence is kept out.

Hearings must regularly be held to discern whether lie detection yet satisfies criteria that might warrant its more routine use, for scientists and polygraph examiners frequently adapt their practices in efforts to gain greater credibility in the criminal justice system. What criteria it must satisfy exactly is the question which presents the biggest challenge to both legal professionals and the experts and entrepreneurs concerned to see the polygraph have its day in court.

US state supreme courts have been inventive in their creation of precedents to support inadmissibility or to control the partial admissibility of lie detection expert testimony. In fact, even when certain criteria can be shown to have been met by proponents of lie detection, the courts have often responded by moving the goalposts and establishing new requirements. Through a century of trial proceedings and opinions they have thereby developed very many ways in which polygraph evidence can be excluded. In general, they do not simply follow *Frye* or not, easily banning or admitting the device. Rather, they have created what I call an 'exclusionary toolkit' – a set of precedents, arguments, opinions, reports and so forth that can be adopted or adapted to the specific questions at issue in any given trial, making for an ever-shifting terrain in which polygraph advocates must grapple for a temporary foothold.

Whilst lawyers and academics have provided ample reports on individual decisions concerning the polygraph's admissibility, they have largely adopted a normative approach, seeking to determine whether these decisions properly reflect broader legal tenets or – in some cases – scientific consensus. Instead of trying to determine whether the decisions were right or wrong, on legal or scientific grounds, I aim to understand the assortment of the many varied decisions, developed across the states, more collectively. What do they comprise as a whole? In this way, I do not offer a legal, scientific or philosophical argument for why lie detection expert testimony should be admissible or not. I prefer to adopt a descriptive approach – to explore the arguments which have been made so far and their effects. In doing so I continue to build an account of the situations in which lie detection features in the criminal justice system, further opening-up the question of uncertainty in the co-production of the polygraph's techno-legal status.

Frye's effect in early cases

Initial polygraph cases in the USA presented a mix of success and failure for its developers. Together these first explorations of lie detection's legal relevance helped forge some of the tools which, in subsequent decades, would be used to exclude the device from criminal courts. Building on existing legal practices for dealing with expert testimony, the first couple of decades of state supreme court cases identified and collected relevant precedents, academic literature and principles, and assessed the broader legal and political commitments that would be refined, revised and extended over the next century.

Of course, the very first case heard on the polygraph, *Frye*, immediately affected how the machine was excluded. One of the key features of the *Frye* approach to scientific evidence was that it maintained a significant degree of legal flexibility, such that it did not prescribe exactly what would count as 'general acceptance' or how courts should determine the 'particular field' to which any given technology, process or principle might belong. Stating that evidence from the polygraph machine might become admissible 'somewhere in this twilight zone' left its advocates in the dark, mostly to the benefit of the court: 'The "general acceptance" standard gave courts leeway for seventy years to decide for themselves whether specific scientific claims were still open to challenge, while all the time ostensibly deferring to science's independent authority' (Jasanoff, 1995: 212).

It was a decade before the next polygraph case was heard before a US state supreme court, this time in Wisconsin. The judges in *State v. Bohner*[2] invoked *Frye*'s ambiguous criteria, which had only just begun to have a broader influence in the US courts' approaches to expert evidence. Whilst it was still relatively new for judges to depend upon *Frye* when determining admissibility of expert testimony, the precedent clearly had a direct bearing on polygraph cases. The evidence against Bohner was substantial enough to convict him of a bank robbery, but he claimed the lower court had erred in not allowing him to submit expert evidence which he claimed would exonerate him of the crime. This time it was Keeler's turn to play the expert, testifying:

> The device gives a continuous quantitative differential blood pressure and pulse curve; that one guilty of a crime becomes disturbed and has distinct emotional disturbances when questioned with reference to the details of a crime which he has committed; that such emotional disturbances are recorded on the 'lie detector' above described.[3]

The judges in *Bohner*, however, agreed with *Frye*, noting that they were 'not satisfied that this instrument . . . has progressed from the experimental to the demonstrable stage'.[4] Despite Keeler's claims that it had been tested over 10,000 times since the 1920s, and a decade of research in lie detection laboratories in Chicago, New York and elsewhere, it seemed Vollmer's disciples had made little dent in the wall of general acceptance. As *Frye* had established: 'Just when a scientific principle or discovery crosses the line between the experimental and demonstrable stages is difficult to define.'[5] The *Bohner* opinion did not trouble to determine where that line would be for Keeler's device, leaving the question open for future courts to decide for themselves, and for polygraph proponents to ponder over.

In addition, the justices in *Bohner* cited the renowned evidence expert John Henry Wigmore (1913), whose brief comments on psychometric testing were first made in his second edition of the *Principles of Judicial Proof*. They were damning. Wigmore had a long-standing disagreement with Hugo Münsterberg, equally celebrated in his own field of psychology and inspiration to various of the early developers of lie detection devices. In a famous debate, Münsterberg complained that the legal system was too slow to adopt psychological facts and techniques, whilst Wigmore contended that the discipline was still in its infancy and had yet to say much of any significance that was not already known to jurists and legal scholars (Magner, 1991).

Five years after this followed two cases in New York. The first, *The People v. Kenny*,[6] opened the door to well-known psychologist and lie detection advocate Father Walter Summers, and the second, *The People v. Forte*,[7] closed it again. It was perhaps the most short-lived period of welcome for the lie detector in any US court. Summers impressed in the first New York case with his testimony to the theory, method and results of his technique but was refused outright in the second.

In *Kenny*, he testified that the use of a pathometer and psychogalvanometer could detect deception with 100 per cent accuracy. The court was convinced and his testimony was admitted:

> According to the testimony of the witness the deductions and the accuracy of the conclusions to be drawn from the examination are undebatable. Both upon legal principle and sound reasoning, it would seem that the courts, if

> willing to accept and receive handwriting testimony, psychiatric testimony and other such expert opinion, should also admit in evidence testimony of the pathometer test and the results disclosed thereby when a proper foundation has been laid therefore.[8]

This marked the first instance in which courts began to compare and contrast polygraphy with other still controversial forensic techniques and sources of expertise. The analyses of handwriting, ballistics, blood spatter and so forth would often be brought into dialogue with lie detection over the coming century. How objectivity, validity and reliability are constituted within various forensic technological processes provide options for courts in making their determinations of the legitimacy of new techniques and the credibility of potential experts. So, although *Frye* understood scientific consensus quite abstractly, in subsequent cases it was not always an abstract notion of 'science' which informed admissibility tests. Instead, legal decision making involves drawing material, technological things together or keeping them apart, showing similarity or difference, generating novel ontological and epistemological expectations depending on which things are brought into consideration and how. The polygraph's reception in criminal trials has been powerfully shaped by its emergence alongside other techniques and technologies. Sometimes to its advantage, and at others to its detriment. In reviewing Summers' research in light of other forensic devices, for example, the New York Supreme Court's opinion was that 'the pathometer' represented a superior approach to discerning witness veracity. But it was its comparison to existing *legal* methods of determining veracity which tipped the scales:

> For hundreds of years our courts have deemed the examination and cross-examination of witnesses in open court to be the best method so far devised for the ascertainment of the truth and have used that method for lack of any better approach. It seems to me that this pathometer and the technique by which it is used indicate a new and more scientific approach to the ascertainment of truth in legal investigations.[9]

The pathometer evidence appeared to confirm that Raymond Kenny was being truthful when he said that he was not guilty of robbery. The jury found him to be innocent. Summers' science had won over at least one state supreme court justice. In his written opinion, Judge Colder reported on several of Summers' laboratory studies. At this stage, his largest study involved 271 individuals and deployed a staged theft protocol (of $20) in which participants were asked to act either as culprits, accomplices or innocents. The relation between Kenny's crime, charged with his second offence of robbery in the first degree, and the laboratory experiment was serendipity, for it helped to establish the direct relevance of the data to the case at hand. But it was some forty-nine examples of use in police investigations, ranging from abductions to murders, which most impressed Judge Colder. Summers claimed that in these cases the results of examinations of suspects had been subsequently confirmed by trial outcomes, demonstrating that the technique was 100 per cent accurate in 'realistic' situations. Introducing polygraph–jury verdict correlations became a key technique in advocates' attempts to have the evidence admitted, and one which would help spur trial courts into developing new arguments for exclusion.

The district attorney prosecuting Kenny also seemed to have been impressed by the results of the pathometer, but he claimed that Summers' research had not yet gained general acceptance and – further drawing on *Bohner*'s interpretation of *Frye* – that it remained in the experimental rather than demonstrable stages of technical innovation. This time the argument failed, and surprisingly it was partly on the basis of Wigmore's *Principles* that Judge Colder reached his decision. Wigmore's brief comments on the current status of the psychogalvanometer were not encouraging but he had claimed that in theory at least, 'If there ever is devised a [reliable] psychological test for the evaluation of witnesses, the law will run to meet it' (Wigmore, 1923). Based on Father Summers' testimony to the accuracy of his device, and due in large part to the strategy of presenting polygraph–jury verdict correlation evidence, Judge Colder was given to believe that time had come.

The news was received with far from unanimous approval. In fact, even balanced reports at the time were troubled by the implications of the decision:

> It must not be forgotten that Judge Colden, though sitting in a criminal case, has, in effect, opened the door leading toward the logical abolition of the jury in all similar cases. If an allegedly scientific fact-finding agency can ascertain the workings of a witness' mind, why require a jury to ponder over the facts when here is an instrument ready-made for such purpose? All those actions, civil or criminal, where the mind is involved, are immediately taken from the jury.
>
> (Forkosch, 1939: 221)

A survey of the jury found that ten of the twelve jurors would still have found Kenny not guilty without the lie detection evidence. This seemed reassuring. More troubling, however, was the finding that five of the jurors were so impressed with the lie detector that they had accepted the testimony without question and four said that they would do so again if they were called upon to serve on a future jury (Forkosch, 1939).

But New York's admission of the polygraph was more a drop in the ocean than it was a sea change. Later that year the state's supreme court rejected the use of the pathometer in *Forte*:

> The record is devoid of evidence tending to show a general scientific recognition that the pathometer possesses efficacy. Evidence relating to handwriting, finger printing and ballistics is recognized by experts as possessing such value that reasonable certainty can follow from tests. Until such a fact, if it

> be a fact, is demonstrated by qualified experts in respect to the "lie detector," we cannot hold as matter of law that error was committed in refusing to allow defendant to experiment with it.[10]

Excluding the polygraph just months after it had been admitted, the New York court's decisions evidence the ambiguous legal and scientific status of the device in these early decades. Again, it was by comparison with other forensic tools, but on this occasion their arrangement and comparison was detrimental to Summers' ambitions. Uncertainty and confidence in the status of other technologies can therefore be brought out differently in different cases, proving a powerful means through which to make sense of novel claims to technological efficacy and relevance. Nonetheless, *Frye*'s emphasis on general acceptance was the most potent factor at work in the period, shaping how new developments in lie detection techniques were understood, how admissibility was made (temporarily) possible, and how exclusions were carried out.

Frye had helped raise a barrier to the devices but it had not resolved the issue entirely. Nor did *Frye* slow the growth in the use of lie detectors within policing practices. Testimony from polygraph experts like Keeler and Father Summers shows that the polygraph had already spread across the US through police investigations. As attorneys continued their efforts to push the lie detector into trials, the courts regularly had to respond to new and recurrent questions regarding the admissibility of such evidence.

Over the years that followed, polygraph advocates in the scientific and legal communities forged ahead. They continued to refine existing theories and adopted new techniques. The machine itself changed significantly as psychologists moved into this area of forensics research and as key figures battled over whose device and technique was most valid and reliable, and how profits should be reaped. Although it risks a naïve reading of socio-technical processes, historians have shown in broad terms that whilst some hunkered down in the laboratory and sought to produce more robust experimental evidence that might more closely adhere to developing forensic science norms and thus convince trial judges, others turned to other markets and emphasised commercialisation (Alder, 2007; Bunn, 2012; Littlefield, 2011). Chief amongst the latter, Keeler patented a machine and began to sell it to a small range of consumers immediately outside of the criminal trial system, including police units and legal firms. Over time, he profited more from training than he did from selling the machine, and his standards slipped. He convinced private investigators, entrepreneurs and others to pay fees to be trained as bona fide polygraph operators (Alder, 1998). A raft of such people operating outside of the trial system then sold their expertise to banks, government departments and a host of medium to large businesses to determine credibility in cases of employment, minor thefts, internal investigations, infidelity, divorce and any number of other scenarios in which trust was at stake. The result was a massive spread in the use of lie detection in the United

States, so that its early struggles to gain acceptance in the criminal trial system were, in entrepreneurial terms at least, made merely a temporary setback.

This expansion also had consequences in the law. The US criminal trial faced an ever-increasing number of requests to admit lie detection evidence in defense and prosecution cases because more and more people were taking the test as part of consultation with their own legal counsel, during police investigations or because they had seen it in magazines, books and on the television (Littlefield, 2011). If the polygraph had lost the early battle inside the courts on the basis of *Frye*, it had certainly won outside them. Criminal trial judges have thereby been forced to continually adjudicate cases over the past seven or eight decades on essentially the same issue: has the polygraph developed sufficiently in the preceding years to now warrant admission in the case at hand? They continue to hear cases in which one party seeks to admit expert testimony as to lie detection evidence and the other requests that such evidence be dismissed. In doing so, they must attend to the specifics of the case, of the proffer of evidence and of state and federal precedents. As such, the state courts have amassed a variety of tools to keep polygraph evidence out of the criminal jury trial, which I term the 'exclusionary toolkit'.

The exclusionary toolkit

Following *Frye*, court strategies for dealing with questions of polygraph admissibility began to diversify. Comparison and contrast with other emerging or more established technologies continued in strategic fashion, but a range of tools were created to respond to changes in the use and development of lie detection outside of the courts and in its immediate vicinity. I present a selection of items from the toolkit so as to build towards an appreciation of how it is and why it is that the courts have often determined to exclude polygraph evidence and expert testimony from a given criminal trial, despite innovations in its form and application, and in distinction to its ever-growing success in other contexts.

The tools I present are:

1 Evidence insufficient scientific agreement;
2 Account for human and technological agency;
3 Argue for practical limitations;
4 Protect the jury.

I do not present all the arguments which courts use in the corpus of more than 200 state supreme court cases that I have closely examined. The selection helps evidence both the breadth of forms but also how they cohere around a key problem for courts regarding lie detection: whether or not the machine risks undermining the way in which criminal trials arrive at their conclusive judgements.

1. Evidence insufficient scientific agreement

The *Frye* decision meant that early cases involving lie detection technologies scrutinised the existing scientific consensus, of which there was little. In the 1930s, *Bohner* and *Forte* had already determined that fifteen or so years of research had done little to improve the machine's standing in the psychology and physiology communities. By the 1940s several state supreme court had reviewed the scientific evidence and heard testimony to the polygraph's status in *voir dire* and in a number of criminal trials. There was thus good reason to refuse it access, for even those scientists most closely knit together by the (still developing) crime detection laboratories found it difficult to agree on basic principles, material arrangement and techniques of use.

Nonetheless, this decade witnessed changes in lie detection law, as gradually more cases were heard across the US, producing key decisions for the future legal status of the machine – for example, the trend of allowing 'prior-stipulation', developed first in Wisconsin (*Le Fevre vs. The State*, 1943)[11] and then in California (*People v. Houser*, 1948).[12] As prior-stipulation shows, the courts were starting to open up but were still generally inclined to reject the device. Cases in Michigan[13] and Georgia[14] characterise the speed with which the polygraph could be dismissed in this decade by use of the 'insufficient agreement' tool. For example, from Michigan, the court in *People v. Becker* simply observed: 'There was no testimony offered which would indicate that there is at this time a general scientific recognition of such tests.'[15] Clearly, attorneys were not all able or ready to mount a case for why polygraph science had advanced since *Frye.*

Creating scientific agreement was indeed a challenge for the polygraph community, for by this time Keeler and Larson had already gone their separate ways, acknowledging that they might never agree on the proper principles for its operation. Moreover, other respected advocates at the time tended to take the position that it was sensible for the courts not to admit lie detection results at trial without prior-stipulation. Fred E. Inbau, colleague of Leonarde Keeler at Northwestern University, was one such expert. In an article in the Boston University Law Review, Inbau (1946) argued that significant advances had been made in polygraph science but that appellate courts' rejections were thus far understandable, if unadventurous.

Cases in Kansas[16] and Nebraska,[17] coming towards the end of the 1940s, show that new legal tools were beginning to be created, and though 'insufficient scientific agreement' remained most prominent, it too was changing. As the polygraph machine spread across the USA into novel contexts and became more celebrated through television and film, criminal cases considering admissibility grew more numerous. Rather than continuing to dismiss cases in which there was no prior-stipulation with a rapid demonstration of 'insufficient agreement', the courts began to demand more of the lie detection apparatus and techniques, requiring scientists and examiners to demonstrate that significant advances had been made in the period since *Frye*, or since their last review of its status. By the 1960s, at least some of the state courts were tiring of the seemingly endless barrage of attempts to get the evidence admitted without prior-stipulation, and – against the more general pattern of maintaining ambiguity – began to set out requirements for change:

> at this time it seems wise to demand greater standardization of the instrument, technique and examiner qualifications and the endorsement by a larger segment of the psychology and physiology branches of science before permitting general use of lie-detector evidence in court.
>
> (*State v. Valdez*, 1962)[18]

Such requirements were difficult to meet, for the psychology and physiology community more broadly was rather sceptical. It was only those who worked directly with the machine who tended to support its use in court. Failing to convince the wider professional community, advocates and examiners responded to these challenges by creating more training courses and forming their own professional organisation, the American Polygraph Association, to try to control qualifications and practices, though with little success. By then the machine had spread far and wide, with no one keeping track of whose hands it had fallen into, on whom it was used or to what ends it was deployed. More problematic still, by virtue of how the device had spread in the US, examiners brought before the courts were more likely to be (former) police officers, and thus scientific novices, than they were to be experts in the physiological sciences. As such, the legal debates regarding scientific agreement on the validity and reliability of polygraph evidence continued well into the later decades of the twentieth century despite a half-hearted professional reform movement. After so long a period of scientific turmoil, fundamental doubts about whether it would ever be possible to give credence to lie detection experts had begun to take root:

> The point is that, given the complete absence of reliable data on base rates, we have no way of assessing the probative value of the polygraph test. Under one set of assumptions, a failed test has some significance, while a passed test does not; under another, the situation is reversed. The figures are further muddied when one recalls that the sensitivity and specificity of the polygraph are also hotly debated.
>
> (*State v. Porter*, 1997)[19]

In fact, some had simply given up waiting:

> [O]ur hope that polygraphy would mature to the point of general scientific acceptance has not materialized. Further hope or expectation in that regard is no longer warranted.
>
> (*Commonwealth v. Mendes*, 1989)[20]

Daubert began to play more of a role in criminal evidence law in some states, meaning that *Frye*'s emphasis on scientific agreement could be diluted. But even here, the *Daubert* focus on reliability was more about trustworthiness than it was repeatability in the traditional scientific sense, meaning that a form of scientific agreement still remained prominent (Gross and Mnookin, 2003: 150). Although the Federal Rules of Evidence were supported in *Daubert*, which in some states' rules had the effect of relaxing the bar on polygraphy, it largely helped sustain exclusion on the basis of community disputes.

Despite the back and forth of attempts to demonstrate or undermine agreement, and even in the face of changes made in the operation of the device, the organisation of the examiner community and in legal practices, the 'insufficient scientific agreement' tool has remained one of the central modes of excluding polygraph evidence from *Frye* to this day. Where *Frye* has framed a court's interpretation of the polygraph, admission without prior-stipulation has generally been denied on grounds of disagreement in other communities, for example in physiology or psychology.

Perhaps the most important feature of this strategy – and why I consider it a strategy rather than a merely programmatic application of *Frye* and/or *Daubert* – is that it is malleable. By preserving a greater measure of indeterminacy in the rules of exclusion, the courts maintain greater control over the interpretation of 'scientific agreement' or 'reliability', making it possible to adjust expectations of the polygraph community as developments occur in their science or in their social organisation. Adapting this through the comparison of polygraph techniques with other technologies or forensic processes adds further room for interpretative flexibility.

2. *Account for human and technological agency*

By the 1950s the courts were becoming more familiar with the polygraph machine. Indeed, one witness testifying to his experience in lie detection claimed to have presented expert testimony at over 2,400 trials in which suspects had confessed following a test (see *State v. Lowry*, 1947).[21] Through experience courts had begun to build a broader set of tools for dealing with requests to admit lie detector evidence, or for knocking back appeals based on claims of admissibility errors. Although demonstrating insufficient scientific agreement remained a core strategy, statements regarding the agency involved in lie detection assemblages became increasingly prominent, affecting and reflecting how the nature of polygraph examinations was understood. This strategy worked by differently highlighting the agency of actors involved in the examination, with different effects, tailored to the specificity of the evidence before the court.

First, courts began to draw attention to the agency of the polygraph machine itself. To market the device successfully in the USA and to convince suspects that it worked, scientists and police officers using the machine would drastically exaggerate its powers. This often crept into expert testimony at trial. For example,

Father Summers had gone as far as to claim his pathometer device was 100 per cent accurate when testifying during the *Kenny* case, and it was not uncommon for experts on polygraphy to assert rates of 95 per cent and above. As hyperbolic claims became commonplace in legal proceedings, the courts began to draw attention to the fact that scientists were more or less arguing that the machine could read minds perfectly. This was, of course, hokum, and it was easily challenged by highlighting epistemological uncertainty evidenced in the academic literature. The exaggerations also had the effect of drawing too much attention to technological agency. On the basis of examiners' claims to near perfection, judges argued that these professed powers were too great for admission to the courtroom, for they meant that the polygraph constituted a witness who, no matter how impartial it might be, could not be cross-examined:

> the machine itself – conceding the comparatively high percentage record as to accuracy and reliability claimed for it – escapes all cross-examination.
>
> (*Boeche v. State*, 1949)[22]

Or, as the state supreme court of Kansas put it:

> The practical effect of the admission of this [polygraph] testimony was to constitute a mechanical device – as reported by the operator – a sort of witness *in absentia* on the question of the defendant's guilt or innocence.
>
> (*State v. Lowry*, 1947)[23]

Such a witness to the defendant's guilt or innocence, immune to the law's own methods of veracity determination, could not be tolerated. Courts that highlighted mechanical agency were thereby able to exclude the device for the very reasons which advocates felt warranted its admission.

Just as the courts could easily constitute the polygraph machine as an active but inscrutable participant in the test, so too could they make it a passive thing, merely serving the interests of its master. The complement to the polygraph as witness was the examiner as judge and jury. As the courts learnt more about how polygraph technology worked and how the machines were actually being used in police stations, they increasingly drew attention to the examiner's role:

> The "lie detector" cannot detect a lie; it is the examiner who must analyze the subject's responses to the "relevant questions" in comparison with the "control questions" in order to determine physiological changes indicative of lying. The validity of the test results depends to a great extent, therefore, upon the expertise of the examiner.
>
> (*Wynn v. State*, 1982)[24]

The examiner's central position in the test was linked to other concerns about the polygraph's misuses, for by the 1970s courts had become aware of the ways in which the lie detection exam could form part of abusive interrogations, meaning that their expectation of scientific agreement, standards of operation and so forth simply escalated. Iowa's state supreme court brought these factors together in one of its key analyses:

> We are also cognizant of the fact the validity of the polygraph result depends on the qualifications, skill and integrity of the polygraph operator. The instrument by its nature is subject to abuse. . . . Furthermore, the breadth, sensitivity and importance of the inference from polygraph evidence demands a higher standard of trustworthiness than is required of other kinds of scientific evidence. . . . In a field where so much depends on subjective analysis by the machine operator and the outcome of trial may well turn on the polygrapher's opinion, we believe a strong showing of scientific acceptance and evidentiary reliability must be made.
>
> (*State v. Connor*, 1976)[25]

Highlighting the examiner's role had begun to lead easily to constructions of the lie detection technique as a merely subjective process, no better than any human mode of determining veracity, seen clearly in the following Connecticut case:

> Although there has been a growing trend among law enforcement personnel toward the use of the polygraph as an investigative aid, its reliability is still subject to question. It is a subjective rather than objective examination, the results of which are influenced by a number of human variables.
>
> (*State v. Mitchell*, 1975)[26]

Against the courts' use of this strategy, the polygraph experts and attorneys struggled to get evidence admitted to trial: claims about the machine's potency led to challenges which emphasised mechanical agency; claims to the examiner's credentials and skill led to challenges which emphasised human agency. The advocates found that they could not win the battle for admissibility when the criminal trial courts were able to adjust their constructions of human and technological agency in this fashion. The courts played off the uncertainties which accompany polygraphy against protestations to perfection or professionalism.

3. Argue for practical limitations

Constraints on legal procedure also began to figure in opinions on polygraph admissibility, as several problems arose in the conduct of hearings on lie detection, which were exacerbated by the increasing frequency of its submission to courts. First, the state supreme courts found that one effect of placing an emphasis on the examiner's role in lie detection tests was to draw attention to examiners' trustworthiness. Courts could thus argue that examiners testifying to their judgement on the veracity of witnesses could themselves become subject to claims of duplicity:

> we can foresee conditions where to ascertain the truth, it would become necessary to require the operator of the machine to submit to a test to determine the truthfulness of his interpretations.
>
> (*Henderson v. State*, 1951)[27]

If the lie detector was going to be used to determine one witness's veracity, then why not that of all of them? As courts began to use this tool, they linked it to problems in the orderly and timely administration of justice caused by the need to constantly update themselves on developments in polygraph science and professional community organisation. Since the advocates of polygraphy continued to make changes in their practices and in the material arrangement of the device itself (for example, adding new measurement components, or, in later years, developing software), new appeals to reliability had to be heard. This meant that polygraph experts were regularly testifying to their credentials and on the machine's reliability, only for another expert to be brought before the court to testify that the device in question did not work and that the method adopted was unreliable. To dispense with the seemingly endless review of the machine's status and the role of the examiner in the process, the courts needed a new exclusionary tool, and so began to argue for practical limitations. For example, by the 1970s some of the state supreme courts had grown impatient with one expert claiming the machine was practically infallible and another claiming it hardly worked at all. It was, the courts argued, a 'battle of experts' that could not be sustained at the expense of justice. This technique sat neatly alongside the 'human and technological agency' exclusionary strategy:

> The orderly administration of justice is of significance to all of us, including defendants. Introduction of polygraph tests, even at post-trial hearings, would produce another trial. The polygraph operator would be the person tried. His (or her) qualifications, his conduct, his conclusions would be disputed. This is a natural result as the test results depend upon the operator's technique, intuition and judgment.
>
> (*People v. Barbara*, 1975)[28]

Placing an emphasis on human agency in the polygraph examination meant that it was important not only to check the credentials of any given examiner but to explore in detail how examiners conducted the test on this occasion. The courts could thus argue that this was unfeasible, for there are limits to how much time can be expended during a trial and there was not enough to conduct a 'trial of the polygraph' on every occasion:

> We are forced to conclude that the administration of justice simply cannot, and should not, tolerate the incredible burdens involved in the process of ensuring that a polygraph examination has been properly administered. If a trial court were to adequately police the reliability of stipulated results,

> the time required to explore the innumerable factors which could affect the accuracy of a particular test would be incalculable.
>
> (*State v. Grier*, 1983)[29]

The state supreme courts have spent decades continuing to make this argument in order to keep pre-trial hearings on polygraphy to a manageable volume in lower courts and to ward off appeals on the basis of exclusion. This tool evidences the ways in which the courts responded to changing social, technical and political forces in their development of the exclusionary toolkit.

4. Protect the jury

The 'practical limitations' tool has, in part, become an effective item in the toolkit, for it can also be used in conjunction with one of the most powerful of strategies for fending off the lie detector: the use of the 'protect the jury' argument. Since the end of trial by ordeal, the modern jury trial had developed incrementally, with the role and responsibilities of the jury gradually taking the form that they do today. The jury trial took the shape that it did in part because of political forces which placed community norms at the heart of justice. Although it helped the courts to adjust to a legal system in which God would play a less active role, worries about the jury have run in parallel with its development, and it is difficult to find a period in which the jury trial has not been criticised by at least some in the legal community. The key difficulty is that even though its epistemological powers have been questioned, the use of a criminal jury has been an important way of ensuring that the determination of guilt is seen to be socially and publicly just, even if most cases are not resolved by its use.

The jury's status as a faulty but necessary mechanism for arriving at the truth has framed testimony on forensic technology from as early as 1930, in the context of fingerprinting, for example (Cole, 1998, 2008). Courts have excluded expert forensic testimony on the ground that it challenged the jury's purview as the final arbiter of the ultimate issue: whether the defendant is guilty or innocent. In the case of lie detection this has been more acute, for the machine's explicit purpose is to determine whether someone is lying or telling the truth when professing innocence. The courts have frequently argued that to allow such expert testimony would be to usurp the role of the jury – however fallible it might be.

> [T]he admission of the results of polygraph examinations would tend to distort the truth-finding process. The danger [is] that the jury will be overawed by the polygraph examiner's opinion.
>
> (*Hinton v. State*, 1989)[30]

The most common form of this argument is for courts to use polygraph advocates' exuberant claims of accuracy to argue that the device is so powerful that the jurors would divest themselves of their responsibility to arrive at an opinion and would instead defer to the polygraph expert's judgement. Ironically, the lie detection community too uses an argument regarding the jury's weakness, but to support admissibility, making it difficult for the community to succeed in such efforts, for the argument tends only to weigh more powerfully against admission.

Fears of deference are, in fact, one issue which has exacerbated the trouble regarding the jury's powers to judge objectively, ever since expert testimony developed. Legal scholarship has, over the decades, proposed a number of remedies for the difficulty of deference, which can be broadly summarised thus: (1) having judges instead of juries decide cases; (2) using 'blue ribbon' juries of highly educated citizens; (3) implementing trial format innovations of various kinds (Hans, 2007: 21). Whilst the jury has been challenged in this way, it has tended to retain its prized place in the public heart of American justice even as it is repeatedly questioned, indicating that something more than objectivity is being preserved. Although most cases are not determined by jury trial, the protect-the-jury argument is one of the most prominent items in the toolkit, for the jury's symbolic power, not only within legal culture, but within American culture more broadly, is even greater than that of the polygraph. Alongside the democratic vote, and other such idealised practices, the jury trial is a quintessential expression, in American ideology, popular culture and its social imaginary, of rule of law, fairness and self-government. As Papke (1999) argues:

> It grows out of American culture in general, and it feeds back into it. [. . .] [W]hile drawing from our sentiments, values, and ideologies, the courtroom trial convention reinforces, shapes and directs. One cannot measure the impact with any precision, but the pop cultural convention likely affects lawyers' attitudes; the public's view of lawyers, courts, and society; and even American consciousness in the most general sense.

Scientific evidence has become a central trope in the cultural forms and genres which centre on or make use of the jury trial. It is hard to think of an episode of *Law and Order*, *The Good Wife*, or *How to Get Away with Murder*, for example, which does not make explicit reference to or does not involve the looming plot presence of science and technology. Legal professionals might not be immune to this sensationalism. Some evidence scholars have argued that when dealing with science, the 'law reifies expert testimony as "evidence" with "weight"' and that by doing so it 'fails to view expert testimony as argument, the ultimate success of which rests primarily on the persuasive power of the expert's *ethos* – the perception of credibility' and not on its actual scientific merit (Scallen and Wiethoff, 1997: 1144). In the context of the polygraph, however, the protect-the-jury argument demonstrates the court's deep understanding of such issues. Indeed, the application of this tool oftentimes depends upon the court's appreciation of how the lie regarding the polygraph's reliability has been propagated through mass media.

Opinions occasionally cite the 'aura of infallibility' (*Scheffer*) which accompanies such testimony as being an important factor in determining to exclude it:

> The questionable accuracy of polygraph tests combined with their aura of scientific reliability has led this court to follow the traditional rule rendering such evidence inadmissible.
>
> (*State v. Esposito*, 1996)[31]

Evidence must be not only credible, but measured and balanced. Cross-examination plays with scientific argument and credibility, requiring performances of reason, judgement and humility (Jasanoff, 1995). Polygraph examiners and scientists struggle in this regard because the machine's basic application relies on a performance of God-like mind-reading capacity (for more on this, see Chapter 6). Ironically enough, playing-up their powers renders them impotent, for as they seek to climb the wall of admissibility they necessarily challenge the jury's purview.

Conversely, when polygraph experts attempt to avoid the protect-the-jury argument by presenting a more modest account of the science of the machine's operation, some courts simply state that the machine can then only serve to provide one more opinion alongside that of the jury and can thus be safely dispensed with. In the face of these complex expressions of expertise, other courts have argued that it is beyond the capacity of jurors to determine how much weight should safely be given to polygraph evidence.

Over the decades courts have thus built these varying applications of the tool into a powerful and flexible argument for exclusion. It was, for example, a key factor in *United States v. Scheffer*:

> Unlike other expert witnesses who testify about factual matters outside the jurors' knowledge, such as the analysis of fingerprints, ballistics, or DNA found at a crime scene, a polygraph expert can supply the jury only with another opinion, in addition to its own, about whether the witness was telling the truth. . . . Jurisdictions, in promulgating rules of evidence, may legitimately be concerned about the risk that juries will give excessive weight to the opinions of a polygrapher, clothed as they are in scientific expertise and at times offering, as in respondent's case, a conclusion about the ultimate issue in the trial.[32]

The importance of protecting the jury from the temptation of deference reveals an implicit conception of how we should justly arrive at a decision regarding veracity, which is especially evident in these opinions but which runs through all considerations of scientific expert testimony. As Roberts and Zuckerman (2004: 319) explain:

> Expert opinion addressing the elements of the offence is sometimes rightly said to amount to a usurpation of the court's, and more particularly the

> jury's, fact-finding jurisdiction. But this is not because there is a comprehensive objection to disputed issues being settled, to all intents and purposes, by the testimony of one witness. This is precisely the role played by key witnesses of fact in criminal trials every day. The real problem in certain cases is that final judgement will not be reached according to normative standards determined by the jury. . . . It is always a moral question whether expert technical standards should take precedence over community norms, as represented in the courtroom by the common sense of a lay jury.

The 'protect the jury' exclusionary tool is thereby an example of an argument found more broadly in legal opinions regarding expert testimony, which evidences the law's need to ensure public acceptability through the jury trial. But there is more happening in the case of lie detection because the polygraph's proximity to the ultimate issue threatens not only to displace the jury function, but also risks opening up the black box of the jury to more scrutiny than it can tolerate. The jury has become for the modern trial what the hand of God was in the trial by ordeal: an idealised part of American identity, and an inscrutable decision-making machine from which there is no recourse: 'The jury's hidden decision making process and its one- or two-word verdicts leave all mistakes and causes for criticism locked in the black box of the jury room' (Fisher, 1997: 706). The jury proved the perfect response to the clergy's withdrawal from trial by ordeal, for it had the same power to determine outcomes in the absence of other forms of proof, and overcame the problem of public failures of sworn testimony, for 'although two oaths all too easily could conflict, the jury's verdict stood alone and, at least within the system's formal bounds, was almost immune from contradiction' (Fisher, 1997: 704–705). As Jasanoff (1998: 733) argues regarding the modern trial, the 'institutional genius of the courts is their capacity to deliver binding solutions' that are acceptable as right not only in the factual sense but in the moral sense also. The jury is the court's magic lie detector, and so it cannot stomach another, for it threatens to expose the sleight of hand on which that justice already depends, the way in which the jury takes all the contested evidence available, chopped-up and thrown into doubt by cross-examination and battles of expertise, but returns with only a single fact. The mistake which proponents of polygraphy make when arguing for the objectivity of lie detection in place of the subjectivity of the jury is that the power of the jury is not its validity or reliability as a social lie detector but its moral power, to determine a truth which is just, not simply correct. Historians of polygraphy have rightly demonstrated the significance of the polygraph's role in popular culture and of the power of popular culture in the lie detector's spread across the US and its recalcitrance in the face of scientific criticism. I add to this that the machine is not alone in this regard. The jury too is woven into the fabric of the American consciousness, in some contexts threaded together with the lie detector in a complementary socio-political fashion, and in others clashing and unflatteringly, too different and yet too similar to be knit.

Conclusion

There is no general pattern of exclusion in the US state supreme courts. Responding to repeated attempts to gain admissibility to the criminal trial, they have produced a diverse set of reasons for excluding polygraph evidence. There are changes in the kinds of reasons given for exclusion over time; as new challenges are made against *Frye* on the grounds of scientific developments, changes in the polygraph process and community and because of specificities of the case at hand, the courts adapt. The arguments and processes outlined above, amongst others, form an exclusionary toolkit, to readily facilitate the creation of just grounds for exclusion in a sensible and oftentimes swift fashion, but still on a case-by-case basis, even as the forms of challenge shift. Where one tool does not work by virtue of the case at hand, another is applied or a new one developed. The result is that the criminal trial state supreme courts are able to use the exclusionary toolkit to continually adjust what is expected of polygraph expert testimony, responding to but also maintaining a degree of uncertainty in polygraph science and its legal reception. The exclusionary strategies adopted in the trial courts are co-produced with the admissibility strategies developed outside of them, generating a kind of polygraph arms race. The key feature of this co-production is uncertainty regarding the techno-legal status of lie detection evidence. This has effects not only in the criminal trial (as I explore in Chapter 5) but also (as I show in Chapter 6) in the police interrogation.

Notes

1 *People v. Zackowitz*, 254 N.Y. 192, 172 N.E. 466, 467 (1930).
2 *State v. Bohner*, 210 Wis. 651, 246 N.W. 314 (1933).
3 Ibid., 318.
4 Ibid., 317.
5 *Frye v. United States*, 293 F. 1013, 1014 (D.C. Cir. 1923).
6 *The People of the State of New York v. Raymond Kenny*, 167 Misc. 51, 3 N.Y.S.2d 348 (1938).
7 *The People of the State of New York v. Vincent Forte*, 279 N.Y. 204, 18 N.E.2d 31 (1938).
8 *The People of the State of New York v. Raymond Kenny*, 167 Misc. 51, 3 N.Y.S.2d 348, 351 (1938).
9 Ibid., 351.
10 *The People of the State of New York v. Vincent Forte*, 279 N.Y. 204, 18 N.E.2d 31, 32 (1938).
11 *LeFevre v. State of Wisconsin*, 242 Wis. 416, 8 N.W.2d 288 (1943).
12 *The People v. William Jack Houser*, 85 Cal. App. 2d 686, 193 P. 2d. 937 (1948).
13 *People v. Becker*, 300 Mich. 562, 2 N.W.2d 503 (1942).
14 *Refoule v. Ellis et al.*, 74 F. Supp. 336 (Ga. 1947).
15 *People v. Becker*, 300 Mich. 562, 2 N.W.2d 503, 505 (1942).
16 *State of Kansas v. Lowry*, 163 Kan. 622, 185 P. 2d 147 (1947).
17 *Boeche v. State of Nebraska*, 151 Neb. 368, 37 N.W.2d 593 (1949).
18 *State of Arizona v. Fernando E. Valdez*, 91 Ariz. 274, 371 P. 2d 894, 898 (1962).

19 *State of Connecticut v. Christian E. Porter*, 241 Conn. 57, 698 A.2d 739, 768 (1997).
20 *Commonwealth v. Benjamin Mendes*, 406 Mass. 201, 547 N.E.2d 35, 41 (1989).
21 *State of Kansas v. Lowry*, 163 Kan. 622, 185 P. 2d 147 (1947).
22 *Eileen Boeche v. State of Nebraska*, 151 Neb. 368, 37 N.W.2d 593, 597 (1949).
23 *State of Kansas v. Melvin Lowry*, 163 Kan. 622, 185 P. 2d 147, 150 (1947).
24 *Henry Joe Wynn v. State of Alabama*, 423 So. 2d 294, 298 (1982).
25 *State of Iowa v. Atwell Junior Connor*, 241 N.W.2d 447 (Iowa 1976).
26 *State of Connecticut v. Robert Lee Mitchell*, 169 Conn. 161, 362 A.2d 808 (1975).
27 *Henderson v. State*, 94 Okla. Crim. 45, 230 P. 2d 495 (1951).
28 *People v. Barbara*, 400 Mich. 352, 255 N.W.2d 171 (1975).
29 *State of North Carolina v. Charles Allen Grier*, 307 N.C. 628, 300 S.E.2d 351, 359 (1983).
30 *Ex parte Anthony Ray Hinton; Re: Anthony Ray Hinton v. State*, 548 So. 2d 562, 569 (Ala. 1989).
31 *State of Connecticut v. James Esposito*, 235 Conn. 802, 670 A.2d 301, 316 (1996).
32 *United States v. Scheffer*, 523 U.S. 303, 313–314 (1998).

References

Alder, K. (1998) To Tell the Truth: The Polygraph Exam and the Marketing of American Expertise, *Historical Reflections* 24 (3): 487–525.

Alder, K. (2007) *The Lie Detectors: The History of an American Obsession*, New York: Free Press.

Bunn, G.C. (2012) *The Truth Machine: A Social History of the Lie Detector*, Baltimore, MD: Johns Hopkins University Press.

Cole, S. (1998) Witnessing Identification: Latent Fingerprinting Evidence and Expert Knowledge, *Social Studies of Science* 28 (5–6): 687–712.

Cole, S.A. (2008) The "Opinionization" of Fingerprint Evidence, *BioSocieties* 3 (1): 105–113.

Fisher, G. (1997) The Jury's Rise as Lie Detector, *Yale Law Journal* 107 (3): 575–713.

Forkosch, M.D. (1939) The Lie Detector and the Courts, *New York University Quarterly Law Review* 16: 202 231.

Gross, S.R. and Mnookin, J.L. (2003) Expert Information and Expert Evidence: A Preliminary Taxonomy, *Seton Hall Law Review* 34 (1): 141–189.

Hans, V.P. (2007) Judges, Juries, and Scientific Evidence, *Journal of Law and Policy* 16 (1): 19–46.

Inbau, F.E. (1946) The Lie-Detector, *Boston University Law Review* 26: 264–271.

Jasanoff, S. (1995) *Science at the Bar: Law, Science, and Technology in America*, Cambridge, MA: Harvard University Press.

Jasanoff, S. (1998) The Eye of Everyman: Witnessing DNA in the Simpson Trial, *Social Studies of Science* 28 (5/6): 713–740.

Littlefield, M.M. (2011) *The Lying Brain: Lie Detection in Science and Science Fiction*, Ann Arbor: University of Michigan Press.

Magner, E.S. (1991) Wigmore Confronts Münsterberg: Present Relevance of a Classic Debate, *Sydney Law Review* 13 (2): 121–137.

Papke, D.R. (1999) Conventional Wisdom: The Courtroom Trial in American Popular Culture, *Marquette Law Review* 82: 471–489.

Roberts, P. and Zuckerman, A. (2004) *Criminal Evidence*, Oxford: Oxford University Press.

Scallen, E. and Wiethoff, W.E. (1997) The Ethos of Expert Witnesses: Confusing the Admissibility, Sufficiency and Credibility of Expert Testimony, *Hastings Law Journal* 49: 1143–1168.

Wigmore, J.H. (1913) *The Principles of Judicial Proof as Given by Logic, Psychology, and General Experience, and Illustrated in Judicial Trials* (2nd Edition), Boston, MA: Little, Brown and Company.

Wigmore, J.H. (1923) *A Treatise on the Anglo-American System of Evidence in Trials at Common Law: Including the Statutes and Judicial Decisions of All Jurisdictions of the United States and Canada* (2nd Edition), Boston, MA: Little, Brown and Company.

Chapter 5

Polygraph uncertainties in the law

> Polygraph examiners contend that conscious efforts to deceive by a rational individual cause the sympathetic branch of the autonomic nervous system to respond and produce a number of involuntary physiological responses. . . . It is not the act of lying per se which causes physiological changes, but the physiological stress created by lying which causes the autonomic nervous system to respond involuntarily. The physiological changes are identical to those which result from the exposure of an individual to a novel situation, or from emotional strain due to fear, anger, elation, excitement, anguish, or other emotion. Therefore, by attaching mechanical devices to a subject's body, a polygraph does not "detect" lies, but only monitors and measures certain physiological functions of the subject.
>
> (*People v. Anderson*, 1981)[1]

By the 1980s many United States criminal courts had become quite used to dealing with polygraph evidence. They selected easily amongst the tools at their disposal to exclude or admit the results according to the specificities of the case, their general position on the *per se* ban and whether they endorsed prior-stipulation. For example, in the case quoted above, *People v. Anderson*, the Supreme Court of Colorado sought to determine whether the inferior court had erred in its trial of the defendant, who had been acquitted for aggravated robbery and first-degree sexual assault. The prosecution claimed the trial court should not have admitted the results of a polygraph exam conducted on Anderson. Expert testimony as to the machine's function, validity and reliability was given at the trial, but neither party had stipulated to the admissibility of the results in advance of the test. To complicate things, Anderson had taken three separate tests by three different examiners. The first was requested by the prosecution and conducted at the local police station, but this proved inconclusive. Anderson then hired his own examiner, who concluded that he was being truthful in his denial of the assault. The district attorney wanted an independent test and ordered a further operator to strap up Anderson to the machine. Truthful again, it concluded, and the defendant filed a motion to admit the results. Determining that the trial court's decision to admit the polygraph evidence had been an error and that the polygraph was inadmissible *per se* in Colorado, the state's supreme court sent Anderson back for a re-trial. Although the court cited *Frye* in support of its decision (see Chapter 3), determining that there was insufficient agreement in the scientific community, Justice Erickson also took time to set out a sophisticated explication of some of the epistemological and ontological foundations, uncertainties and unknowns in polygraph science.

There have been numerous such cases in the past century in which nuanced and well-resourced opinions have been written, constituting a long history of legal analysis on the merits and detractions of the science and practice of polygraphy, regarding exactly how the machine works, what can and cannot be known from its operation, and how it is used within policing and in the courtroom. This sophisticated analysis has drawn upon principles, precedents and reasoning, but also on a range of scientific studies, exhaustive expert testimony and lengthy government investigations, stirring together a mix of scientific and legal ways of knowing into a techno-legal hodgepodge.

The expert in agony

This kind of mixture is well known to science and technology studies researchers, who have identified a range of forensic devices and practices which have been co-produced at the nexus of science and law. In the process of such co-productions, legal accounts sometimes idealise scientific work. This is partly because of the image of science which its practitioners themselves work to sustain, as a purely objective revelation of nature (Mnookin, 2007). But it is also strategic: in accounts of justice and expert testimony it is not uncommon for judges and attorneys to adopt naïve forms of realism and representations of scientific work in order to sustain their own claims to legitimacy (Jasanoff, 1995; Wynne, 2013). Expert evidence, in this regard, can help to play a role in the maintenance of legal truth regimes, for it supports claims to impartiality in the administration of justice. The idealised images of science sometimes incorporated into legal rules and opinions on expert testimony help nurture a self-image of legal practices as similarly dispassionate in the application of technology in the service of truth (Mercer, 2002).

It could thus be said that scientific uncertainty is anathema to criminal law, for how could justice be carried out where the facts remain in question? But it is precisely within situations of uncertainty that legal practices have developed for managing justice with imperfect knowledge, making criminal trials, much like laboratories, hotspots of uncertainty work. As Star (1985) outlines, there are multiple forms of uncertainty which occur in scientific work, deriving from – for example – efforts to standardise practices, generalise findings, and manage everyday and institutional politics. As with scientific spaces, legal spaces might also involve the resolution, management and amplification of uncertainties. The criminal court is characteristic in this regard, not least when it comes to the testimony of expert witnesses.

As such, a common finding in studies of forensic technologies is that what once seemed certain and unquestionable in the laboratory can once again become contested, appear uncertain and be subject to interpretation in the criminal court (Cambrosio et al., 1990). Uncertainty, complexity and multiplicity must be negotiated, and can be exacerbated, by the ways in which technical objects move from one context to another, or from one legal regime to another (Cloatre, 2015).

Exemplifying this broader phenomenon, the adversarial system in United States criminal courts pits opposing parties against each other as they each seek to make their own case as solid as possible, building evidence piece by piece, whilst demolishing the opposition's reasoning and evidence step by step. Jasanoff (2008: 126) adapts Latour and Woolgar's (1986) notion of 'agonistic fields', spaces in which scientific claims are challenged, adapted, repaired, dismissed or concretised, for use in describing the criminal trial. She argues that the adversarial system is an agonistic field of contestation in which are debated competing representations of expertise, scientific techniques and the evidence which they provide. As I outlined in Chapter 4, showing that something is uncertain is a key tactic used by attorneys in their development of a case and in their cross-examination of witnesses, which is reflected in court decision making. It brings doubt to the facts and thereby opens reasonable grounds for acquittal or opportunity to dismiss a defence. This shapes the court's approach to the efficacy of science. Although idealised images are common, in other ways courts are primed to see science as a product of human action.

Cross-examination also introduces uncertainty into accounts of science by the nature of the turn taking system of talk on which it depends. The question and answer approach allows witnesses to speak but constrains what they can say. Speakers jostle with each other, attack and defend, thrust and parry, in the construction of a shared account. They pre-empt each other's talk, producing a co-constructed line of argument (and rebuttal) which leaves some things difficult to pin down and others more suddenly concluded (Drew, 1992).

Through this process, the expert can become the most visible mediator of scientific facts. Rather than using the hesitant, questioning, situated and embodied way which they might within their offices, laboratories or lecture halls (Latour, 2010), they must instead respond within the to-and-fro of the adversary system, answering questions both for the defence and the prosecution, attempting to present the facts but – as often as not – wandering into territory upon which they are not directly qualified to speak. This is part of a more general issue with expert testimony: scientists' expertise is also on trial (Cole, 2002; Lynch and Cole, 2005; Lynch et al., 2008; Cole, 2008) and they can find "themselves manipulated as mere tools in the hands of the lawyers" (Golan, 2004: 54).

As Lynch argues, through his analysis of DNA admissibility and the work of the O. J. Simpson 'Dream Team', the 'general acceptance' standard for admissibility, as inaugurated in *Frye*, when combined with the adversary processes of the law, privileges a kind of objectivity made up of contestation, deconstruction, construction and webbed interactions (Lynch, 1998: 854). Experts testifying before a court must, through sometimes quite protracted means, demonstrate their expertise in *voir dire.* This means showing how they acquired their knowledge, through education and experience, and providing warrant for their claim to speak on behalf of certain facts. Their objects and theories, devices and data must thereby be described, understood and critically assessed through trial practices to service the case and public justice more broadly (Lynch, 2007). The result is that legal uncertainty work does not necessarily align with laboratory, clinical or scientific modes of managing unknowns or of conferring confidence.

Since scientific evidence can be a crucial part of an attorney's case, expert testimony has become an industry. The industry which has grown around expert testimony has further complicated the way in which uncertainty is managed in trials. There are companies and registers that provide links between attorneys and possible experts, as well as experts who – on making more money in trials than they ever could in academia – determine to make a career of being expert witnesses. A conflict with the idealised image of science as dispassionate is inevitable when scientists make knowledge for increasingly large consultancy fees. This has at least coincided with, and perhaps directly caused, an increasing suspicion of such evidence, leading to a supposed crisis in expert testimony to which *Daubert* and the Federal Rules of Evidence were supposed to be the answer (Jasanoff, 1995; Haack, 2009). But these and other key rules and decisions have not solved the problem of credence in expertise because the production of uncertainty regarding their merit is inherent to the adversarial process, rather than being only a product of financial interests, much to the displeasure of some involved (Haack, 2007). The result is a lingering concern about charlatanism, for it can be exceedingly difficult to pin down who is a ratified expert and who is a fraud (Roberts, 1999).

The lines between intentional misrepresentation and incompetence is even more difficult to draw and harder to police, but is something that must be resolved in practice on a regular basis. In this way, the law regularly deals with lying, deception, fabrication, omission and ignorance in scientific talk. Innumerable variants of half-truths and overextensions of the facts must be explored to find the guilty but also to protect the innocent in a legally and permanently binding fashion. In an upending of the techno-legal history of lie detection, it is scientists who are put on the rack.

If the process of presenting and challenging the facts in cross-examination is set up to expose uncertainties on both sides, then one key purpose of the judge and jury is to make decisions on the basis of such uncertainty. In the closure of legal proceedings certainty becomes paramount, as Brian Wynne (2013: 143) argues, because 'the need to explain decisions as if they were absolutely certain is bound up with the problem of maintaining social authority.' In some degree, every trial of an individual party is also a performance of social order. Judges write opinions for an 'imagined public', as Lynch (2016) puts it, as though the public were hovering over their shoulder. As such, when the courts engage with scientific evidence and claims about reality, they are not merely concerned with how reliable this information is, but also with what such information means for the court's modes of producing certainty, and how it might support or detract from its obligations to enact public forms of justice. Whether expert testimony can be admitted and what role it plays at trial depend partly upon its potential effects and uses within this particular legal magic.

It is through this process of exposing contingencies to conjure legal certainty that polygraph science must be articulated, applied or rejected. Claims regarding the intention to deceive and its bodily effects must be resolved for the law, not merely for science. When determining whether polygraph evidence is admissible, the courts will, on a regular basis, produce accounts of the polygraph machine, how it works and on what key assumptions it depends.

How such uncertainties are handled in a given state will be affected by the legacy of cases in that state and how the courts draw upon precedents set elsewhere; how they use scientific evidence, government reports and so forth as part of trial processes; on the resources they have to hand in the exclusionary toolkit; as well as on their more ephemeral techno-political temperament. In this chapter I explore the way in which the Supreme Judicial Court of Massachusetts has managed its apprehension of uncertainty at the heart of polygraph examinations.

The polygraph provides an excellent example of how ontological claims must be understood as situated within practices and with regard to specific norms and politics (Lynch, 2013). The court opinion is a specific form of producing ontological statements about socio-technical relations.

To explore this issue further I examine how ambiguities about the role of polygraph operators (their status as experts or not) and uncertainties in the polygraph machine's operations (regarding the relationship between the intention to deceive and bodily responses) were managed in Massachusetts over a thirty-year period. Massachusetts is an exemplary state for such an investigation: the series of cases from 1963 to 1993 evidence the complex relations between uncertainties in science and in criminal justice.

Ontological uncertainty in the state of Massachusetts

The story of the polygraph's status in the Supreme Judicial Court of Massachusetts began in 1963, a few months before the assassination of President John F. Kennedy. As a native of the state, Kennedy had served Massachusetts for thirteen years, first in the House of Representatives and then in the Senate before winning his presidential campaign in 1960. Coming towards the end of what many thought would be the first of two terms, he gave some of his most famous speeches on peace, civil rights and the terrifying prospect of nuclear war with the Soviet Union.

Deep racial, sexual and gender divisions in American society had pervaded the decades following World War II and generated much cause for revolution and political change (Gitlin, 1993). The assassinations of John and then Robert Kennedy, Malcolm X and Martin Luther King made for a bloody and frightening period of disunity. These 'long years of the 1960s' have since been characterised as a period of 'civil war' in the USA (Isserman and Kazin, 2004). And there is no doubt that wars at home were matched by wars abroad, as conflicts with the USSR, Cuba and Vietnam intensified (Lytle, 2005).

The 1960s was a period of immense change, both within the states and in relations with other countries. Amongst the seething political ferment, technologies of various kinds had their roles to play. Questions of social responsibility loomed large for the scientists of the great American engineering projects in nuclear weaponry and space exploration (Wisnioski, 2012). These were not just science as usual, but rather a series of investments in national defence and the economy, set against a developing politics of globalisation. American national identity was increasingly tied to Soviet national identity, such that the 'space race' and other global competitions came to power these nations' 'socio-technical imaginaries' (Jasanoff and Kim, 2009; Pickersgill, 2011). Techno-politics were being reconfigured as a result of these various changes in engineering, nationhood and war.

However, a period of quite unprecedented optimism about the technological future preceded a gradual shift from trust to suspicion, from hope to fear:

> To be an American engineer in the aftermath of World War II had been to look upon a seemingly limitless future. . . . In the United States, the period from 1945 to 1964 was one of near-utopian belief in technology's beneficence. Few denied that the nation was undergoing a scientific revolution, as popular imagery portrayed a futurist world of flying cars and plastic houses. . . . In the decade that followed, technology took on ambiguous and ultimately sinister connotations in American thought and culture. . . . [C]ontemporary life was becoming more alienating, more destructive, more totalitarian, and less human.
>
> (Wisnioski, 2012: 3–4)

Technologies which previously held promise and appeared to have solved long-standing problems now began to take on a sinister aura. This was best characterised by the nuclear bomb, but also in other contexts such as in the growing use of agricultural chemicals, like DDT, which seemed to have ushered in a new economic era for American farming, only to plunge it into controversy as pollution and environmental degradation followed soon after (Lytle, 2005). But it was probably the Cold War threat that was most influential with regard to technological expansion:

> Much has already been written about the role of scientists in building the nuclear weaponry that defined the Cold War, but the space race, the exploration of the deep oceans and the deep interior of the Earth, the rise of telecommunications and civilian nuclear power, and many other scientific and

> technological developments were also directly tied to the global conflict that the Cold War entailed.
>
> (Oreskes, 2014: 1)

So it was that the post-war period saw the culmination of a series of expansions in polygraph use in the US, which had begun with Senator Joseph McCarthy's 'red scare' and continued as fears about infiltration from Soviet Russia and China continued into the seventies and eighties. Initially fruitless attempts at admissibility in the courtroom were not long mourned as the developers and users of lie detection devices succeeded far beyond their expectations in business and government. A range of employers had begun using the polygraph to screen applicants for jobs or to conduct internal investigations of theft, absenteeism and so on. The dark shadow of risk, doubt and fear cast over the states brought with it the rise of lie detection as an entrepreneurial industry:

> By the middle of the 1960s, some 1000 operators were at work, a tenfold increase since the 1950s; and half of the nation's police departments were making use of the technique, along with thirteen federal agencies, and some 40,000 businesses.
>
> (Alder, 2007: 254)

In an era characterised by suspicion of communist spies and corporate espionage, the polygraph established its place as part of an ongoing war against enemies both foreign and domestic. It was also the point at which the polygraph's inculcation with everyday criminal investigation practices properly took hold. The training of thousands of operators included a great number of police officers who took one of the short courses offered by several entrepreneurs. Chief amongst them was Leonarde Keeler, who had patented his own device and did more than anyone to secure a place for the machine in the heart of American society. Regular use of the polygraph in corporate America and police investigations led to a growing number of criminal cases in which lie detection evidence was submitted for consideration.

Regarding evidence law, the *Frye* decision had spread quite widely across the US and had attracted criticism from key legal experts. The most prominent of evidence scholars, Charles McCormick, for example, argued that whilst *Frye* was an important decision to be used in determining admissibility, it should not be adopted as *the* rule. Rather, it should be taken into account as part of the broader remit to consider '[a]ny relevant conclusions which are supported by a qualified expert witness', which 'should be received unless there are other reasons for exclusion. Particularly, its probative value may be overborne by the familiar dangers of prejudicing or misleading the jury, unfair surprise and undue consumption of time' (McCormick, 1954). McCormick's views represented not just the need to produce truth within criminal trials but to do so within the broader remit of securing a just society in dangerous times. Fears over false convictions, entrapment and police use of psychological tactics were beginning to peak (Leo, 2008). The truth could not come at any price, and so polygraph science had to take its place within these broader principles and aims.

The ambivalent status of *Frye* meant that polygraph expert testimony was reviewed with caution as courts navigated their responsibilities to admit pertinent evidence and keep out hogwash. Court mechanisms for dealing with polygraph evidence were multiplying, meaning that a range of positions across the US developed. They could not easily be synthesised into a coherent framework, and indeed as forensic science expanded and Americans became wary of certain technological developments, messy techno-legal pictures became a common spectacle. This is what eventually led some critics to claim that there was a crisis of junk science in the US courts and that the Federal Rules of Evidence, which were developed in the 1960s and 70s, had not adequately resolved the issues.

In the year before his death, President Kennedy had appointed Byron R. White to the United States Supreme Court. Justice White was seen to be a liberal-leaning judge and served until 1993, when he retired. Shortly before he and his colleagues had passed a unanimous verdict on the case of *Daubert v. Merrell Dow Pharmaceuticals* (see Chapter 3), which introduced a more authoritative, though some would say equally confused, statement on the ways in which expert evidence should be reviewed (Haack, 2005).

In that tumultuous thirty-year period between the first case heard in 1963 and the introduction of *Daubert* in 1993, the Massachusetts Supreme Court heard ninety cases in which reference to the polygraph was made. As it considered the shifting legal and scientific terrain which characterised polygraph evidence, the state's Supreme Judicial Court went back and forth on its position regarding polygraph admissibility. Its decisions were of course made by reference to the specifics of the cases heard but were also informed by broader concerns about technology, engineering and science, set against the backdrop of social and political upheaval, all of which entangled in the legal system with a growing unease with *Frye*'s precedent of 'general acceptance'. As the years passed, American courts heard cases in which they used *Frye* but with greater degrees of anticipation of the Supreme Court's engagement with the issue of expert evidence and the slowly concretising Federal Rules.

Of those ninety cases, five of them – heard between 1963 and 1989 – make up the basic skeleton for a sequence of techno-legal movements: the key steps in the ontological choreography (Cussins, 1996) of Massachusetts' dealings with polygraph evidence. They begin with *Commonwealth v. Fatalo* (1963).[2]

Angelo Fatalo was convicted of assault and battery with a dangerous weapon, having been identified by the victim as one of four men who had attacked him in his apartment. Although Fatalo requested a polygraph test during the police investigation, on this occasion none was administered and it was not until he was in prison following conviction by jury trial that he was given a test to mount an appeal. The superior court denied Fatalo's pre-trial motions to admit the evidence, so he appealed to the Supreme Judicial Court.

The state supreme court applied the *Frye* test of general acceptance, finding that the polygraph had not made any gains in the intervening forty years of scientific work and societal uptake. The court opined that 'there is hardly a device which has caused greater controversy among "experts," lawyers, physicians, psychologists, government officials, and the public in general.'[3] It was 'the controversy swirling around the polygraph test' which evidenced a lack of general acceptability for the Massachusetts court, but besides the worry that scientists could not agree that it worked, what the polygraph *was*, exactly, remained in question. A lie detector? Junk science? An active agent or a passive tool? Unsurprisingly, the court ruled that the denial of Fatalo's motion to admit evidence from his tests was not error, and affirmed his conviction.

Drawing on the academic literature, the opinion also cited a range of 'variables which vitiate the alleged effectiveness of the tests',[4] including: emotional unresponsiveness, ability to beat the machine, physiological abnormalities, mental abnormalities, nervousness and extreme emotional tension. Such variables are all, in one way or another, linked to a body which is fundamentally changeable. That one person's body could be different from another's, by virtue of physiological, psychological, emotional or social factors, made trouble for the key ontological claim at the heart of polygraphy: that the body's responses to acts of deception are regular. If the connection between the body and the 'deceptive consciousness' (to use William Marston's phrasing; see Littlefield, 2011) could be affected by such factors as nervousness or physiological abnormalities, then what would it matter that the machine produced reliable results? That there was disagreement over the reliability of the device was problem enough, but that the device might not even be able to detect lies was disastrous. The court did not develop a fleshed-out account of what a body is and what it does, but it did propose that the account which is to be generally found in most polygraph literature was unsatisfactory. The court excluded the polygraph by raising the spectre of ontological uncertainty. The finality of the court's decision on Fatalo's fate kept open the question of how the polygraph worked and whether lie detection was even possible. Examiners and other experts would now have to provide an account of how the polygraph operated which more definitively secured a valid foundation for lie detection, better connecting the body and the lie, but without any real sense of what would be adequate for the court's purposes.

At the same time, although the Massachusetts court closed the door to the polygraph, it did not lock it, as the opinion concluded: 'We do not hold that such recognition must be universal or that the test must be proven infallible, but rather that the substantial doubts which presently revolve about the polygraph test must be removed.'[5] Implicitly, the court promised to hear the issue again if better evidence could be provided that these problems had been resolved to the agreement of a greater proportion of relevant experts. If, it seemed to state, a more general agreement can be reached on the founding relations between the body and deception, then it would reconsider the matter.

It took a decade before the court was presented with another opportunity to review the status of such claims. The 1974 case *Commonwealth vs. A Juvenile*[6] concerned a thirteen-year-old male who had first been convicted of murder in the second degree of a nine-year-old friend. The evidence was circumstantial – no one had witnessed the crime. Testimony was given, however, that the juvenile knew the victim was dead before anyone else, that there was no reasonable way for him to know this save for having committed the crime, that he knew the boy had been stabbed and the locations of the wounds and that he knew the site of the murder and had searched it for a knife with a friend. Evidence was also admitted that blood found on the defendant's trainers could have been human. The jury found that this proved that the defendant had killed the boy. However, in presenting his case, the defendant had not been allowed to use transcripts of testimony previously given by the witnesses to cross-examine them against their now differing testimony which they gave before the criminal jury. Appealing the decision on this basis, he was afforded a new trial, at which he was found guilty of manslaughter.

It was a case ripe for the polygraph: no direct evidence and a young, vulnerable defendant who claimed he was telling the truth but without any means to prove it. During the second trial the defendant sought to introduce test results from an exam conducted upon him, at his own request, a month after the murder and before his first trial. It was Charles Zimmerman, a trusted polygraph operator for the Boston Police Department, who ran the test. He was well known in the state, for he had famously conducted examinations of the suspects and witnesses in the Boston Strangler case. He now strapped up a thirteen-year-old to the machine. Deception not indicated, so Zimmerman said.

The defendant argued that some recent decisions in other states, which allowed polygraph evidence into trial (mostly with prior-stipulation) meant that there was now precedent which would permit the judge to admit the evidence. He also argued that the field of polygraphy had significantly advanced in the ten years since *Fatalo*. Whilst the motion was denied, the court concluded:

> Since the year 1963, polygraphic testing has become more reliable, and scientific acceptance thereof has increased. Advances have been made both to the machine itself and in the ways of administering the test. When conducted by competent polygraph examiners under proper examination conditions, polygraphic testing possesses a high degree of reliability and scientific acceptability.[7]

The decision thus made an epistemological case for the admission of polygraph evidence – it had become more reliable. However, the judge also made it clear that it was up to a superior court to change the state's procedures for dealing with polygraph evidence and that he could not admit the testimony against precedent.

In hearing on this matter, the Supreme Judicial Court determined that advances had indeed been made in the mechanism of testing and its reliability but that general acceptance had yet to be attained in the relevant scientific community. Nonetheless, rather than upholding the *per se* ban, enacted by *Fatalo*, the court moved towards a prior-stipulation position (see Chapter 3). It would now allow trial judges to admit – at their discretion – polygraph evidence with prior-stipulation so long as it was clearly assured that the defendant's constitutional rights had been protected (e.g. against self-incrimination). Advances in the device's operation, it seemed, warranted a relaxing of the rules against admissibility despite some incredulity in the scientific community. In weighing its responsibility to use only legitimate expert evidence against defendants' right to present their case and marshal what evidence they could, the court was beginning to question *Frye* even though ontological uncertainties remained. The rising use of polygraph examinations in policing, industry and government was beginning to tip the scale towards prior-stipulation.

To justify this new position, the court set out a further ontological account, one that was careful to point out that at least some socio-technical relations in the polygraph's assemblage depended on assumptions rather than demonstrations of fact:

> The polygraph machine is a scientific device which through measurement and recording of involuntary bodily responses – blood pressure, pulse rate, respiration, and skin resistance to electricity – seeks to determine whether an individual is telling the truth. The fundamental assumptions of polygraphy are that (1) there is a regular relationship between lying and emotional states and (2) there is a regular relationship between these emotional states and measurable physiological changes in the body. These assumptions are disproved to a certain extent by the undisputed fact that some persons can tell undetectable lies, e.g., pathological liars, emotionally unresponsive subjects who have rationalized their behavior, children with an active imagination unable to distinguish between reality and fantasy, and others.[8]

The polygraph seemed to measure things that the body was doing, but it remained unclear whether this really bore any causal connection with a lie being formulated. The status of the polygraph machine as a lie detector was again challenged by reference to a body irregular in nature, but also by reference to differences in subjectivity. The court worried that the pathological liar's body would be unresponsive to lies by virtue of repetition and habit; fantasist children's because they themselves do not know what the truth is. Such differences, the court was implying, render the ontological connection between deception and body still uncertain, causing problems for the claims of polygraph experts.

The court also identified a potential avenue through which such ontological uncertainties could be managed: the examiner. Whilst the fundamental assumptions about how polygraph machines work had not changed in the ten years since

Fatalo, what had changed was the organisation of polygraph experts. Key polygraph scientists, entrepreneurs and court decisions were responsible for bringing focus to the examiner as the crucial factor in determining the validity of an examination in order to try to overcome the challenges posed by emerging tools in the exclusionary toolkit (see Chapter 4).

For example, one prominent criminologist, Fred Inbau, had first encountered the polygraph machine when he met Leonarde Keeler in Chicago (Alder, 2007: 117). Inbau wrote influential books and papers on the legal status of lie detection and its use during interrogations, becoming a powerful figure in the community during the decades between the 1940s and 1970s. He was 'a law professor with a difference, a criminal law specialist who was the police officer's friend and ally in the never-ending battle against crime and criminals' (Thomas, 1998). In this guise he was destined to be a friend to August Vollmer's intellectual progeny of scientist cops.

Indeed, his colleague, John Reid, was a Chicago street cop turned polygraph examiner who would go on to develop one of the most widely used interrogation strategies – the Reid technique – which was designed to produce confessions. The company he founded, John E. Reid and Associates Inc., continues to train investigators and polygraph examiners in government, industry and police services around the world. Having previously been reticent to challenge the courts' decisions on lie detection, by the 1960s and 70s, and now working with Reid, Inbau co-authored articles pushing for more legal acceptance of the polygraph machine and chastising the courts for failing to recognise the proper scientific community of lie detection experts. They rebuked 'psychologists who conclude that the lie detector technique is unreliable, without getting out of their armchairs to study and observe the technique as it is being utilized in actual case situations' (Inbau and Reid, 1964: 470).

It was in that same year, however, that the Reid technique was cited as both the most used and the most dangerous mechanism in American police interrogations in the United States Supreme Court decision on *Miranda v. Arizona* (1966).[9] As the *Miranda* opinion states:

> It is obvious that such an interrogation environment is created for no purpose other than to subjugate the individual to the will of his examiner. This atmosphere carries its own badge of intimidation. To be sure, this is not physical intimidation, but it is equally destructive of human dignity.[10]

As *Miranda* spread from Arizona to other American courts it began to shape how confessions were received into evidence (Thomas III and Leo, 2012). So too did the names 'Reid' and 'Inbau' spread as their work and expert testimony began to be called upon in making determinations over polygraph admissibility across the US.

As well as being the year in which *Miranda* was decided, 1966 saw the founding of the American Polygraph Association (APA). Its ambition was to accredit members with professional status and to standardise training programmes for operators. In part, this was a direct response to the challenges posed by court opinions on polygraph testimony, which had often required more 'general acceptance' in a relevant community. It also sought to combat the growing concerns about the role that interrogators could play in producing confessions, characterised and powerfully censured by *Miranda*. By shaping a community of polygraph examiners, the APA hoped to establish a community of practice which might encourage professional conduct but also represent a group which could demonstrate general acceptance of polygraph science. Polygraph examiners speaking before courts began to draw on the APA as a way of establishing their legitimate credentials. As Charles Zimmerman claimed in one instance of testimony, the American Polygraph Association is 'patterned and styled after the American Bar Association and the American Medical Association'.[11] Indeed, it had been developed to mimic professional norms in other communities. It convinced some judges and led the Massachusetts court to instruct the jury that experts are those with superior knowledge, such as 'surgeons, architects, polygraph experts, engineers, ballistic experts, and handwriting experts'. Placing polygraph operators amongst such other more well-established professions evidenced the developing tendency to legitimise their standing and qualifications.

However, both scandals and legal and scientific studies had by now made it clear to scientists, lawyers, judges, polygraph experts and the police that the crucial factor in eliciting confessions during interrogations (with or without the lie detector) was the investigator. When it came to trial, it was ever more important not just to show that the polygraph machine worked, but to demonstrate the superior expertise and sophistication of the examiner. Figures such as Reid, Inbau and Zimmerman, over and above the usual standards of everyday polygraph operators working as policemen or private investigators, sought out a special classification as expert consultants who could be better received by the law. So it was in keeping with shifts in the techno-politics of polygraph expertise that *A Juvenile*, in mounting his defence of murder, proffered that when conducted by especially qualified examiners under appropriate conditions the polygraph was now valid and reliable.

The Massachusetts court was convinced that at least some changes had occurred in training and professionalism of polygraph examiners and that some such superior experts could be trusted to provide scientifically credible and legally relevant opinions:

> Equal to the importance of the machine itself is the competence, experience, and education of the test examiner. . . . Overall, the training and background of polygraph examiners have become more sophisticated in recent years.[12]

But the opinion was not resoundingly positive:

> Nevertheless, as a profession, polygraphy has yet to achieve a uniform minimum level of competence.[13]

There were, nonetheless, signs that key actors like the APA, Reid and Inbau were beginning to have an impact on court opinion regarding general acceptability. Citing Reid and Inbau's (1966) book, *Truth and Deception*, the court felt that polygraph testing had 'come to be regarded as reliable in a growing number of circles, governmental, scientific, legal and private'. These scientists-cum-entrepreneurs were beginning to affect how some courts delineated the relevant community of acceptance. The growing use of the polygraph outside criminal trials was being leveraged to open up the court's doors. This was music to the ears of its advocates. But again, doubt remained regarding fundamental ontological relations:

> There still remains in certain portions of the scientific community resistance to polygraphy's basic assumptions as to the relationship between lying and certain emotional states, and the relationship of those emotional states to measurable physiological changes in the body.[14]

Trying to respond to this jumble of positions, of the uncertainty in the polygraph's basic ontological claims, questions regarding its reliability, shifts in training and professionalisation of examiners, as well as pressure from prominent experts, the court fashioned a fuzzy decision which allowed this uncertainty to be maintained and decisions to be made in face of it. The *A Juvenile* decision left it up to trial judges to find their own position case-by-case on whether the examiner who administered the test lived up to requirements to provide expert testimony. In effect, Massachusetts committed itself to its own experiment with lie detection:

> [P]olygraphy has, for decades, been the subject of study, debate and controversy. It is too late in the day for just another study. Rather, the time is ripe for cautious judicial examination and evaluation. The experience gained may well lead to a total rejection of the concept. On the other hand, it could develop into a very useful tool in the administration of criminal justice. Actual testing in the courts is necessary before this determination can be made.[15]

If the scientific community could not resolve to determine if lie detection worked, then Massachusetts' supreme court would have to step in. By attempting to find a nuanced position amongst competing accounts of polygraph expertise and its basic assumptions, and in trying to do justice to the case at hand, in which a thirteen-year-old boy was convicted of murder via circumstantial evidence, against an expert's opinion that he was being truthful in his denial of the crime, the court provided its own uncertain account of the polygraph machine. In effect, it ruled that the evidence could be admitted – given any individual trial judge's feelings regarding the expert in front of him/her – despite fundamental uncertainties about validity. The broader politics shaping ambivalence towards technology in the 1960s, the development of prior-stipulation since the 1940s, alongside growing political ferment regarding an individual's rights to a fair trial and protection against the more condemnable practices used during police interrogations all shaped the court's decision. It was a co-production of uncertainty at the nexus of technology, law and politics.

In regard to *A Juvenile*'s appeal, the court allowed him to file for a new trial at the superior court, where he could put forward a motion to admit the results of a new polygraph test, to which he would stipulate in advance to the admission of the evidence, whether favourable or not to his case. The defendant did so, but the superior court immediately sent the case to the state supreme court, since the boy was now eighteen years old and the court's jurisdiction over him was therefore in question. Regarding the polygraph evidence, the superior court was of the opinion that *A Juvenile* was no longer a fit subject for polygraph testing. Informed by the operator who had originally conducted the exam in 1970, the court claimed that no meaningful re-examination of the defendant could be conducted. Five years had passed, in which the boy had demonstrated a 'behaviour pattern of unlawful activities',[16] which – in the opinion of the examiner – 'would overshadow' the events about which he would be questioned. In effect, the boy had changed: his memories, bodily responses, beliefs, fears and so forth were no longer those of the boy first questioned. Science and the law now made him a man, and, having served his sentence as a juvenile by this time, he was released without further trial.

In the two years that followed, the experiment with lie detection was put into action. Lower courts encountered a variety of petitions to admit polygraph evidence and reached a range of conclusions as to whether experts could testify, under what circumstances and for what purpose. The legal position was uncertain, which was a direct result of the ambiguities in the *Fatalo* and *A Juvenile* decisions. More clarity was required. The Supreme Judicial Court thus took the opportunity to transfer a case from a court of appeals to its province and determined to clarify the terms under which polygraph evidence could be used. In trying to resolve the issue by bringing *Commonwealth v. Vitello* (1978)[17] up to the state's chief court, Massachusetts was again faced with uncertainty regarding the polygraph's fundamental assumptions about bodies, minds and social action. Drawing on the academic literature, a further refined account of the machine was constructed:

> The polygraph method consists of a complex, and incompletely understood, series of interactions among (1) the subject, (2) the polygraph machine, and (3) the examiner. Each of these three facets of the "lie detection" process can be broken down into constituent parts. Focusing first on the response of the subject, it is theorized that (a) the act of lying or deceiving leads to conscious conflict, (b) which induces fear or anxiety, (c) which results in measurable physiological events such as changes in skin resistance (perspiration), respiration, blood pressure, heart rate, blood flow, skin temperature, muscle tension,

> pupillary diameter, gastric motility, and blood oxygen saturation. . . . These psychophysiologic reactions are largely controlled by the autonomic nervous system, which functions automatically and involuntarily. . . . . The theory that conscious lying can be detected reliably by measuring involuntary physiological changes thus seems to assume a regular relationship between lying and certain emotional states, and a regular relationship between emotional states and changes in the body. The validity of both these assumptions has been questioned.[18]

This was the most detailed description the court had provided. Drawing on its developing knowledge of the technique, existing literature and having heard polygraph experts testify before lower courts during its experimental phase, the Supreme Judicial Court was now well placed to pass judgement on the machine. But it was still the examiner who became the focus of attention:

> The second facet of the polygraph method – the mechanical and electrical functioning of the polygraph machine – appears to be the best understood, and least controversial, aspect of the process. . . . The activities of the examiner in administering and analyzing the polygraph examination constitute the most controversial facet of the polygraph method.[19]

The *Vitello* opinion provides one of the most thoroughgoing analyses in legal decisions in the US of the way in which operators influence the examination and of their significance in making the device work *as* a lie detector. In effect, the decision helped the Massachusetts court to determine that the expert examiner is the agent who helps to produce the connection between mind and body, by virtue of inducing in the individual a fear of being caught and by implanting in the individual's mind the belief that the polygraph is infallible.

What historians have carefully described in the twenty-first century had already been uncovered in 1960s and 1970s Massachusetts Supreme Judicial Court opinions. Major US cases like *Miranda* had sensitised judges to the power of interrogation practices. Shifting techno-legal practices in relation to *Frye* and the Federal Rules of Evidence, as well as political movements in the US more broadly, informed a nuanced approach to the way in which polygraph machines worked when operated as part of police interviews. The courts situated the science.

So it was that in *Vitello* Massachusetts synthesised its previous opinions, critically assessing expert testimony and academic scholarship and making a number of its own inferences and deductions to show that '[s]tatements and actions of the examiner directed at establishing the infallibility of the polygraph process are thus essential prerequisites to eliciting physiological responses of noticeably distinctive character.'[20] The way in which the body and mind responded to each other, measurable by the polygraph, was deeply dependent upon the social process through which the examination was conducted. The court argued:

> A crucial part of the testing phase – that period when the polygraph machine is actually attached to the subject – has nothing to do with the operation

> of the machine. Rather, the examiner is instructed to observe carefully the behavior of the subject to detect tell-tale signs of deception. The work by Reid and Inbau is replete with references to supposed behavioral clues.[21]

The ontological implications were dire:

> First, the polygraph machine and the recordings it produces are merely tools in the hands of the examiner; the machine is not independently capable of discerning truth from deception. . . . Second, the successful use of the polygraph method is highly dependent on the mental attitude and beliefs of the subject at the time the test is administered. If the subject does not believe in the infallibility of the process, or is unconcerned or fatalistic about the outcome, the efficacy of the procedure is seriously compromised. A similar problem may be presented by a subject who has successfully rationalized and justified his actions, or who has cleared his conscience by confessing to a clergyman prior to the test, and who does not respond to the examiner's efforts to arouse the subject's emotions.[22]

This was a dissection not only of how the polygraph was used to make knowledge but of what its object of knowledge was: the relation between the mind and body was not, in the final analysis, regular, because the examiner made the bridge. The court ruled that polygraph evidence could not be used by either the prosecution or the defense as part of their cases in chief, since how it worked was compromised and what it measured was fundamentally uncertain; therefore, reliability of the machine could never be confirmed. Selecting some additional resources from the exclusionary toolkit (concerns about jury deference and the costs of frequent pre-trial motions), the court all but returned Massachusetts to its position in *Fatalo*, fifteen years prior, except that it allowed some discretion in admitting polygraph evidence for the limited effect of evidencing a witness's general credibility. The polygraph was used in Massachusetts to this end for a few years (e.g. *Commonwealth* v. *Moore*, 1979)[23] and rules were gradually refined.

After several increases in restrictions, the court eventually returned the state to a *Frye*-based *per se* ban in 1989, in *Commonwealth v. Mendes*.[24] Mendes was charged with rape of a child by force, indecent assault and battery of a child under sixteen. He motioned the court for a polygraph examination. At the same time, Kenneth M. Rosenberg, also charged with rape of a child, incest and indecent assault and battery on a child under fourteen, motioned the court to admit evidence from a court-ordered polygraph test. An evidentiary hearing in the Superior Court took four days and the judge determined that whilst the polygraph evidence would be admissible on scientific grounds and given established precedent regarding stipulation and so forth, it could no longer be admitted on grounds of policy. It used the 'argue for practical limitations' tool: the time taken to present the evidence, cross-examine the expert witnesses and then go through expert rebuttal was so long that it was 'not worth the price'. The judge also had various questions of law as to how to deal with evidence from previously taken tests as compared with court-ordered tests but also whether the defendant's previous experience with polygraph examinations would invalidate any future test. In effect, the judge asked: does his experience after arrest but before trial change him in such a way as to make it impossible to tell if he is lying or not any longer, and does it make it such that the polygraph will no longer work on him? In responding to this most recent challenge to polygraph admissibility, the Supreme Judicial Court opined, 'We have never determined that the appropriate scientific community, which includes physiologists and psychologists, has generally accepted the validity of polygraphy as a scientific means of detection deception.' In other words, the gains made by polygraph advocates were lost.

Regarding ontological concerns, the opinion provided a further account of how the polygraph works; this time the examiner and the questioning technique were thoroughly interwoven with the description:

> In brief, the polygrapher investigating a criminal incident usually employs the "control question technique." Under this technique, the examiner asks the subject "relevant" questions pertaining directly to the incident being investigated, and also "control" questions which deal with acts similar to the incident in question but are more general in nature. The control questions are designed to produce a strong physical reaction on the polygraph instrument, which records respiratory activity, sweat gland activity, and changes in blood pressure. The examiner asks several sequences of control and relevant questions, and then scores the result using one of several scoring methods. As a general matter, if control questions elicit stronger reactions than the relevant questions, the subject is considered to have answered the relevant question truthfully. If the relevant questions produce stronger responses, then the subject is considered to have been deceptive.[25]

In its first encounter with the polygraph in *Fatalo*, Massachusetts had cited numerous challenges to the connection between the body and the deceptive intention. *Mendes*, twenty-five years later, was the culmination of much uncertainty work regarding the socio-technical and legal status of polygraph examinations. It now presented a version of the apparatus which firmly placed it within the procedures of the exam, which linked the body and mind of the examinee to the questions designed by the examiner to 'produce' a strong physical reaction and clearly placed the determination of the results in the hand of the examiner, who 'scores the result using one of several methods'.

As experts in polygraph examinations had gradually become more prominent, they had responded to early failures in admissibility by professionalising themselves, developing training courses, changing the apparatus and examination procedures, developing new ways of presenting themselves as experts alongside surgeons, fingerprint analysts, ballistics experts and so on. This helped them to challenge the use of the insufficient agreement tool for exclusion but put them right at the heart of the exam. As they took greater control over their role as credible experts, they also became the wedge that unstuck the connection between body and mind. The Massachusetts court, at least, determined that it was the examiner who was essential to joining the deceptive intention to the bodily response through their orchestration of the exam and not any reliable and consistent 'natural' connection between lying and fear.

Unable to extricate themselves and resolve disagreements in the community of practitioners, psychological and physiological experts could too easily be pitted against each other, not only as regards the interpretation of any one chart but as regards the very foundations of polygraph science. The court's selective use of the image of science as objective, dispassionate and a revelation of pre-existing reality could not be aligned with the expert operator manipulating the subject into fear and anxiety, to produce the responses scribbled onto the chart. The amount of uncertainty work required to manage the uncertainties which resulted from the socio-technical and legal status of lie detection was no longer sustainable, and the polygraph was once again ejected from the Massachusetts Supreme Court.

Conclusion

The result of processes of dealing with expert evidence, of ways of dealing with knowledge and uncertainty, is not 'a clash between "hard scientific facts" and legal conventions or representations', but rather 'a back and forth movement between different kinds of representation' (Cambrosio et al., 1990: 277). Rather than a simple binary opposition of science and law, where ontologies might be seen to clash with each other, the co-production of science and law leads to a messier result. Ontological categories and distinctions are mixed up and produce novel ambiguities. Lie detection is no stranger to the result of this transduction.

Whilst scientific research on polygraphy has often assumed a causal connection between the intention to lie and the murmuring of the body, meaning that the body could be listened to for signs of the mind, it has not been so easy to maintain that assumption when the device is proffered for admission to criminal trials. Courts considering the admissibility of lie detection evidence have regularly brought out this assumption and interrogated its soundness, developing a hundred years' worth of critical examinations of the polygraph's ontological foundations. The result is an ontological uncertainty regarding the device.

Rather than face complete rejection or acceptance, lie detection advocates have been caught up in a dance of expertise, whereby they must constantly work to remake polygraph practices to better fit the ever-shifting legal terrain. In doing so, however, they continue to keep the polygraph from being adequately black-boxed for general acceptance. Changes in practices, in the device's operations and in communal norms and expert presentations help entrepreneurs, defendants and their advocates get another chance at admissibility, but also prevent them from succeeding in any consistent fashion over time or across the criminal trial courts in the United States. When connections between the body and the lie are being scrutinised, it is in a situated fashion, with regard to the changing political, legal and technical landscape, meaning that lie detection regularly gets its day in court, but rarely without ontological trouble. In the following chapter, I explore how these uncertainties and the examiner's critical role in the exam are negotiated in the use of the polygraph in police interviews. What poses a significant limitation in the trial court represents a powerful opportunity for manipulation of suspects in the police interrogation room.

Notes

1 *The People of the State of Colorado v. Richard Anderson*, 637 P. 2d 354, 360 (Colo. 1981).
2 *Commonwealth v. Angelo Fatalo*, 346 Mass. 266, 191 N.E.2d 479 (1963).
3 Ibid., 480.
4 Ibid.
5 Ibid., 481.
6 *Commonwealth v. A Juvenile*, 365 Mass. 421, 313 N.E.2d 120 (1974).
7 Ibid., 123.
8 Ibid., 125.
9 *Miranda v. Arizona*, 384 U.S. 436 (1966).
10 Ibid., 457.
11 *Commonwealth v. Thomas J. Foley*, 7 Mass. App. Ct. 608, 612 (1979).
12 *Commonwealth v. A Juvenile*, 365 Mass. 421, 313 N.E.2d 120, 124 (1974).
13 Ibid., 125.
14 Ibid.
15 Ibid., 129.
16 *Commonwealth v. A Juvenile*, 348 N.E.2d 760, 370 Mass. 450, 454 (1976).
17 *Commonwealth v. Peter Vitello*, 376 Mass. 426, 381 N.E.2d 582 (1978).
18 Ibid., 586.
19 Ibid., 587.
20 Ibid., 588.
21 Ibid.
22 Ibid., 590.
23 *Commonwealth v. Moore*, 379 Mass. 106 (1979).
24 *Commonwealth v. Benjamin Mendes*, 406 Mass. 201, 547 N.E.2d 35 (1989).
25 Ibid., 39.

References

Alder, K. (2007) *The Lie Detectors: The History of an American Obsession*, New York, NY: Free Press.

Cambrosio, A., Keating, P. and Mackenzie, M. (1990) Scientific Practice in the Courtroom: The Construction of Sociotechnical Identities in a Biotechnology Patent Dispute, *Social Problems* 37 (3): 275–293.

Cloatre, E. (2015) Fluid Legal Labels and the Circulation of Socio-Technical Objects: The Multiple Lives of Fake Medicines, pp. 97–144 in D. Cowan and D. Wincott (eds.) *Exploring the 'Legal' in Socio-Legal Studies*, London: Palgrave Macmillan.

Cole, S.A. (2002) *Suspect Identities: A History of Fingerprinting and Criminal Identification*, Boston, MA: Harvard University Press.

Cole, S.A. (2008) The "Opinionization" of Fingerprint Evidence, *BioSocieties* 3 (1): 105–113.

Cussins, C. (1996) Ontological Choreography: Agency through Objectification in Infertility Clinics, *Social Studies of Science* 26 (3): 575–610.

Drew, P. (1992) Contested Evidence in Courtroom Cross-Examination: The Case of a Trial for Rape, pp. 470–520 in P. Drew and J. Heritage (eds.) *Talk at Work: Interaction in Institutional Settings*, Cambridge: Cambridge University Press.

Gitlin, T. (1993) *The Sixties: Years of Hope, Days of Rage*, New York, NY: Bantam.

Golan, T. (2004) *Laws of Men and Laws of Nature*, Boston, MA: Harvard University Press.

Haack, S. (2005) Trial and Error: The Supreme Court's Philosophy of Science, *American Journal of Public Health* 95 (S1): S66–S73.

Haack, S. (2007) Of Truth, in Science and in Law, *Brooklyn Law Review* 73 (3): 985–1008.

Haack, S. (2009) Irreconcilable Differences? The Troubled Marriage of Science and Law, *Law and Contemporary Problems* 72 (1): 1–23.

Inbau, F.E. and Reid, J.E. (1964) The Lie-Detector Technique: A Reliable and Valuable Investigative Aid, *American Bar Association Journal* 5 (5): 470–473.

Isserman, M. and Kazin, M. (2004) *America Divided: The Civil War of the 1960s*, Oxford: Oxford University Press.

Jasanoff, S. (1995) *Science at the Bar: Law, Science, and Technology in America*, Cambridge, MA: Harvard University Press.

Jasanoff, S. (2008) Representation and Re-Presentation in Litigation Science, *Environmental Health Perspectives* 116 (1): 123–129.

Jasanoff, S. and Kim, S.-H. (2009) Containing the Atom. Sociotechnical Imaginaries and Nuclear Power in the United States and South Korea, *Minerva* 47 (2): 119–146.

Latour, B. (2010) *The Making of Law: An Ethnography of the Conseil D'état*, Cambridge: Polity Press.

Latour, B. and Woolgar, S. (1986) *Laboratory Life: The Construction of Scientific Facts*, Princeton, NJ: Princeton University Press.

Leo, R.A. (2008) *Police Interrogation and American Justice*, Boston, MA: Harvard University Press.

Littlefield, M.M. (2011) *The Lying Brain: Lie Detection in Science and Science Fiction*, Ann Arbor: University of Michigan Press.

Lynch, M. (1998) The Discursive Production of Uncertainty: The OJ Simpson "Dream Team" and the Sociology of Knowledge Machine, *Social Studies of Science* 28 (5/6): 829–868.

Lynch, M. (2007) Expertise in Action: Presenting and Attacking Expert Evidence in DNA Fingerprinting Cases, *Villanova Law Review* 52: 925–952.

Lynch, M. (2013) Postscript Ontography: Investigating the Production of Things, Deflating Ontology, *Social Studies of Science* 43 (3): 444–462.

Lynch, M. (2016) Preliminary Notes on Judges' Work: The Judge as a Constituent of Courtroom "Hearings", pp. 99–130 in M. Travers and J.F. Manzo (eds.) *Law in Action: Ethnomethodological and Conversational Analytic Approaches to Law*, Abingdon: Routledge.

Lynch, M. and Cole, S. (2005) Science and Technology Studies on Trial: Dilemmas of Expertise, *Social Studies of Science* 35 (2): 269.

Lynch, M., Cole, S.A., Mcnally, R. and Jordan, K. (2008) *Truth Machine: The Contentious History of DNA Fingerprinting*, Chicago, IL: University of Chicago Press.

Lytle, M.H. (2005) *America's Uncivil Wars: The Sixties Era from Elvis to the Fall of Richard Nixon*, Oxford: Oxford University Press.

Mccormick, C.T. (1954) *Handbook of the Law of Evidence*, St. Paul, MN: West Publishing Company.

Mercer, D. (2002) SSK and Law/Science Encounters Involving Controversial Science and Technology: A Brief Critical Overview, pp. 254–272 in A. Beal (ed.) *2002 Yearbook of the Institute of Advanced Studies on Science and Technology*, Graz: Profil.

Mnookin, J.L. (2007) Idealizing Science and Demonizing Experts: An Intellectual History of Expert Evidence, *Villanova Law Review* 52: 763–802.

Oreskes, N. (2014) Introduction, pp. 1–10 in N. Oreskes and J. Krige (eds.) *Science and Technology in the Global Cold War*, Boston, MA: MIT Press.

Pickersgill, M. (2011) Connecting Neuroscience and Law: Anticipatory Discourse and the Role of Sociotechnical Imaginaries, *New Genetics and Society* 30 (1): 27–40.

Reid, J.E. and Inbau, F.E. (1966) *Truth and Deception: The Polygraph ("Lie Detector") Technique*, Philadelphia, PA: Lippincott, Williams and Wilkins.

Roberts, P. (1999) *Tyres with a "Y": An English Perspective on Kuhmo Tire and Its Implications for the Admissibility of Expert Evidence*, International Commentary on Evidence, www.degruyter.com/view/j/ice.2004.1.2/ice.2004.1.2.1005/ice.2004.1.2.1005.xml, last accessed 10/06/2017.

Star, S.L. (1985) Scientific Work and Uncertainty, *Social Studies of Science* 15 (3): 391–427.

Thomas, R.M. (1998) Fred Inbau, 89, Criminologist Who Perfected Interrogation, *The New York Times*, May 28th, https://www.nytimes.com/1998/05/28/us/fred-inbau-89-criminologist-who-perfected-interrogation.html, last accessed 06/07/2017.

Thomas III, G.C. and Leo, R.A. (2012) *Confessions of Guilt: From Torture to Miranda and Beyond*, Oxford: Oxford University Press.

Wisnioski, M. (2012) *Engineers for Change: Competing Visions of Technology in 1960s America*, Boston, MA: MIT Press.

Wynne, B. (2013) *Rationality and Ritual: Participation and Exclusion in Nuclear Decision-Making*, Abingdon: Routledge (Earthscan).

Chapter 6

Polygraph interrogations

> "I knew I couldn't lie. I should have just refused to take this test, but how does that look. That's just as bad, right?"
>
> (*People v. Storm*, 2002)[1]

The polygraph played a central role in an appeal against a conviction of homicide in the case of *People v. Charles Storm*. The appeal case hung on the degree to which the defendant had been lied to and manipulated by the police. The polygraph had been pivotal in producing his confession because he had at first denied any involvement in the murder. However, after he had agreed to take a lie detector test, the police were able to lead him towards a confession. Once the exam was over, the operator said the polygraph showed a greater than 99 per cent chance that Charles was lying about his role in his wife's death. Although he would later plead guilty to the homicide, the polygraph first helped the police extract a confession of assisted suicide. Faced with the seeming objectivity of the lie detector results, Storm claimed his wife had begged him to help her to end her life, and that after much resistance he had finally agreed to her request. Only now, having said too much, did he realise that he should have refused the test: 'She told me not to tell anyone, but here I am telling [you].'[2] Without legal counsel, when invited by the police to take a polygraph, Storm had found himself unable to refuse, because, as he said, 'that [looks] just as bad'. The conceit that refusal indicates guilt was a powerful influence in the construction and use of 'polygraph talk' (see Chapter 3) but is nowhere more forceful than in police interrogations. Storm's appeal was denied and his conviction affirmed. But not all cases in which confessions are produced through deceptive use of the polygraph are so easily swept aside. Some produce false confessions. In this chapter, I explore the mechanisms of polygraph interrogations and show how the deceptions upon which they operate can result in deeply problematic outcomes.

Although the specific forms of physical brutality which constituted torture as a form of interrogation are no longer commonplace in the pursuit of criminal investigations, Lynch notes:

> Today's interrogators . . . have 'gentler' means at their disposal for extracting confessions from subjects who would prefer to give less damning accounts of their actions.
>
> (Lynch, 1998: 854)

Polygraph tests represent one such 'gentler' means, for they are regularly used during interrogations to produce confessions. However, even in states where the machine is *per se* banned from admissibility, polygraph tests can have a powerful effect in a trial as well. Indeed, they might serve as pivotal trial evidence, as many states allow polygraph-derived confessions to be admitted without mention of the device. This is a problem, and is especially dangerous in cases involving false confessions.

Alder (2007: 22–23) identifies the examiner as an expert in the intimidation and manipulation of suspects, referring to them as 'consummate antiprofessionals'. However, his analysis is based primarily on the work of Keeler and others rather than on the ways in which the machine is used routinely by 'ordinary' examiners in criminal cases. To develop my analysis, I expand on Alder's contention, rooting my investigation of how the polygraph works in empirical examples of police interrogations of criminal suspects to show exactly how it is that confessions are produced through use of the polygraph machine. Although social histories such as Alder's have shown how the public image of the lie detector serves to help produce confessions, there remains very little evidence of the conversational and practical means through which examiners use this public image to draw out information from subjects, or of other techniques made possible by the lie detector.

To develop this account, I use five interrogation transcripts (four of which have been anonymised), including the full transcript of the interrogation of Peter Reilly, used with permission and in great detail in the final section. The first section of the chapter provides a broad overview of how lie detection and police lying have been connected, drawing on existing social histories of interrogation practices and the polygraph machine. The middle section outlines four techniques of interrogations (performing objectivity; performing expertise; epistemological and ontological certainty; and bleeding, clearing and composing), which I develop based on empirical data. The final section uses this analysis to show how these techniques led to a false confession in the case of Peter Reilly, exemplifying the power and danger of polygraphy.

Police lying, the polygraph and confession

It is acceptable in American law for the police to lie to suspects under certain circumstances. In fact, some interview techniques support deception, for lying is only 'proscribed, on a case-by-case basis, only when it violates a fundamental conception of fairness or is the product of egregious police misconduct' (Skolnick and Leo, 1992: 5). The comfort with deception within policing practices (Hunt and Manning, 1991) grew alongside the gradual decline of torture leading up to the 1900s, and following the rise and fall of the third degree in the first half of the twentieth century (see Chapter 2).

Deception is tolerated and sometimes encouraged to the extent that it helps to produce confessions. The professional attitude to confession has swung back and forth in the legal history of the USA, having at various times and by various commentators been treated with deep scepticism or lauded as the gold standard of evidence. For most of the 1800s and early 1900s, American law followed the 'Hawkins–Leach dictum' (Thomas III and Leo, 2012), that confessions induced by hope or fear, in however small measure, were to be treated with great suspicion. It was the decline of the third degree that resulted in a shift towards acceptance of confessions drawn out through deception. Whilst the professionalisation of the police force and the rise of forensic technology did not therefore reduce the number of confessions extracted through questionable methods, it did mean they could be produced without – for the most part – resort to physical brutality.

Alongside this change, new legislation and policing and pre-trial processes were created to improve the effectiveness of the courts and prosecution service, which were importantly framed by an emphasis on increasing the reliability but also the admissibility of confession evidence (Bryan, 1997). In effect, many of the techniques and strategies that developed around investigation and interrogation were designed to produce confessions in the absence of outright torture. As the polygraph expanded across the United States in the middle of the twentieth century, American courts were becoming more comfortable with, and indeed more dependent upon, the use of confession. Key to this shift was the development of *Miranda* (see Thomas and Leo, 2012 for a thorough review), which sought to ensure that the interrogation techniques in operation were not physically abusive and that the accused was adequately informed of his/her rights and warned against self-incrimination. But the co-production of criminal trial law and police practices has led to a situation in which a great many confessions are wrung from criminals without physical inducement but with powerful psychological and social force. Police lying and manipulation have been at the heart of this development, and it should be no surprise that the polygraph machine has fit snugly alongside these methods, for it too depends upon a lie.

The development of police deceptions, tricks and manipulations was refined during the professionalisation of the police force in the first few decades of the 1900s. It was August Vollmer's campaign for scientific and humane policing (Alder, 2007; Bunn, 2012) which – perhaps unexpectedly – made one of the most significant contributions to the emergence of new interrogation procedures that would ultimately be founded upon lying. The polygraph machine, in John Larson and Leonarde Keeler's hands, was supposed to be the means of bringing about positive changes to investigatory practices; to allow policemen to leave their clubs and coshes behind. Besides the moral crusade, there was clear epistemological reasoning behind the campaign because Larson – much like the lawyers and judges of the 1800s – knew that certain forms of interrogation could produce confessions not only of varying quality but also of varying truthfulness. People might confess to a crime they did not commit:

> By this procedure one favorite method of eliciting the truth is for a relay of detectives to quiz the suspect for a period of hours during which he is given

> no food or allowed to sleep until a confession is extorted. Naturally with the suspect in an extremely fatigued condition the nature of the confession will vary according to the suggestibility of the subject and the suggestions of the examiner.
>
> (Larson, 1925: 225)

Today it is much better established that certain coercive interview techniques and interrogation strategies currently in use can bring about false confessions (Lassiter, 2006). But even in the 1920s the group of lie detection experts developing the polygraph knew that suggestion could lead to false confessions, and they had high hopes that the lie detector could help eliminate the need for these kinds of tools. If the suspect might not always speak truthfully under physical or psychological duress, then perhaps the physical itself could be made to speak by means of mechanical measurement. The lie detector's moral and scientific rationale was thereby premised on interrogations being conducted civilly and without brutality. However, the use of scientific techniques did not displace other emerging and shifting interrogation strategies but instead became tangled up with them, such that today the polygraph is regularly used to support other coercive tools, which can affect how the examination is conducted.

Once John Reid and Fred Inbau had taken on the mantle from Larson and Keeler in the 1950s, as advocates for, and refiners of, the polygraph test (Inbau and Reid, 1964), the irony of the lie detection exam was doubled, for as they vastly expanded the role of polygraphy they allied it with the use of new techniques of deception in US interrogation practices. Rather than 'civilising' policing in the scientific tradition, they handed detectives around the country a new weapon with which to extract confessions by advocating a particular approach to interviewing which would ensure that the polygraph's function was far from objective and more in keeping with the growing use of psychological duress.

The police interview or interrogation is almost always a 'closed social interaction' (Irving and Hilgendorf, 1980), meaning that there is a concomitant social pressure on the interviewee, isolated in an unfamiliar room, seated a little too near their accusers, involved in a conversational pattern which is quite unlike everyday talk and lacking any real control over the process. The space lends itself to techniques which leverage these micro-power relations.

Playing on these relations, Reid developed practices for aligning lie detection with the interrogation process, including the widely used 'control question technique'. Alongside which, he designed a range of 'behaviour provoking' questions which, proponents still believe, induce subjects to reveal their true thoughts through involuntary physical behaviours (Inbau et al., 2013). Eventually, Reid consolidated his various innovations in lie detection and interrogation into the eponymous 'Reid technique' of interviewing criminal suspects.

The Reid technique rests upon and makes use of a certain idea about the body, which was already well integrated with polygraph science, and had been crucial to practices of torture: the notion that the body reveals underlying truths involuntarily and automatically if manipulated in the correct way. This assumption has, over the past century, become commonplace not only in policing, but in psychology, psychiatry and neuroscience, permeating cultural media and inducing in many an expectation that our bodies will betray us. Reid's technique takes advantage of this fear, heightening it to encourage people to betray themselves, not through body spasms but through speech. It is designed to provoke seemingly voluntary confessions by guiding the body's psychological tensions towards confession. As one critic points out:

> Comprised of nine steps, this approach is explicitly designed to induce admissions of guilt by increasing the anxiety associated with denial while minimizing the anxiety associated with confession.
>
> (Kassin et al., 2010: 42)

The Reid technique encourages police lies, deceptions, tricks and so on, for it understands them to be indispensable ingredients of such interrogations: 'The vast majority of criminal offenders are reluctant to confess and must be psychologically persuaded to do so, and unavoidably by interrogation procedures involving elements of trickery or deceit' (Sear and Williamson, 1999: 73).

Reid's approach does not have to depend upon a lie detector but often uses one. It also encourages police officers to become lie detectors themselves, using supposedly observable physiological cues to deception to hone their questions and challenge suspects' claims. In practice, both the machine and the Reid strategy are used in tandem. And much like the polygraph, the Reid technique has often been questioned on the grounds of validity and reliability, with studies showing that its fundamental assumptions are problematic and its application far from perfect (Kassin et al., 2010), which is especially troubling given its status as the most widely used police interrogation approach in the United States.

So although the polygraph has always depended upon trickery and chicanery to convince subjects of its powers (Alder, 2007; Bunn, 2012; Littlefield, 2011), in the 1950s it started to become an assistant to Reid's recommended investigatory strategies, compounding the use of lies to compel the truth. The result is that '[o]ver the last fifty to sixty years, the methods, strategies and consciousness of American police interrogators have been transformed: psychological persuasion and manipulation have replaced physical coercion as the most salient and defining features of contemporary police interrogation' (Skolnick and Leo, 1992: 3). The polygraph successfully spread across the United States for a range of reasons, but it was its role in helping navigate the transition from bodily torture to psychological manipulation which perhaps best served its successful bid to become an integral part of American policing. Ironically, the spread of police lying helped spread the lie detector.

But how exactly do lie detection and police lies go hand-in-hand? As Skolnick and Leo (1992: 7) report in their review of interrogation techniques in the USA, 'one of the most common physical evidence ploys is to have the suspect take a lie-detector test and regardless of the results – which are scientifically unreliable and invalid in any event – inform the suspect that the polygraph confirms his guilt.' From out of this situation a confession might be wrought for any number of reasons. The truly devious element in the Reid technique's partnership with lie detectors is that its final interrogatory step of 'turning an oral confession into a written confession' involves the removal of any reference to having taken, or having refused to take, a polygraph examination from the content of the confession (Inbau et al., 2013: 186). This is to prevent the confession from being rejected as evidence during a criminal trial, because polygraphs are of course often banned as evidence, and their mere mention within criminal trials where there is no prior-stipulation can result in grounds for appeal (see Chapter 3's consideration of 'polygraph talk'). It is therefore a sensible measure from the perspective of criminal prosecution. But it also erases an oftentimes-crucial part of how a confession was arrived at. The exclusionary toolkit, developed to refuse the polygraph admissibility in an ever-shifting socio-technical landscape, also has the effect of encouraging investigators to rewrite the story of the confessions they have elicited. Exclusion of the polygraph from many trial courts has resulted not in an attenuation of its powers, but in the intensification of its ability to influence justice from behind the scenes.

There is a tacit acknowledgement of this ambivalence regarding the lie detector's status in the justice system. Several states which uphold the *per se* ban on the polygraph retain a knowing ignorance of the machine's usefulness within police practices despite its lack of scientific credentials, for example in *State v. John Kolander* (1952):

> We have no doubt that the lie detector is valuable in investigative work of law enforcement agencies and may frequently lead to confessions or the discovery of facts which may ultimately lead to the solution of many crimes. But we are in accord with the rule that the lie detector has not yet attained such scientific and psychological accuracy, nor its operators such sureness of interpretation of results shown therefrom, as to justify submission thereof to a jury as evidence of the guilt or innocence of a person accused of a crime.[3]

In allowing for this backstage role, the courts too frequently ignore the fact that the person accused of a crime might only find himself in front of a jury because of a confession brought about through use of the lie detector in deeply troubling ways. Preventing the proper exploration of the lie detector's deceptive role in any given criminal investigation serves not to protect the jury from the machine but to hide it from them, and with sometimes dire consequences. As Alder (2002: 17) puts it, 'the machinery of American law thrives on the (potentially) coercive exercise of discretionary power.'

The manipulative techniques of polygraph interrogations

In the middle section of this chapter I outline a number of techniques used by polygraph interrogators to extract confessions from suspects, drawing on key interrogation transcripts. Since the analysis depends on an understanding of the ways in which polygraph exams are conducted, it is worth briefly reminding the reader of some key points. First, the polygraph exam takes the following format: pre-test, test phase, post-test. The pre-test is ostensibly used to construct the exact questions to be asked in the test phase; the test phase is when the questions are asked whilst the subject is strapped up to the polygraph machine; and the post-test comprises an interview with the subject about the results. However, as will become clear in the analysis, the lines between these phases are intentionally blurred to manipulate the subject. Equally as important is the blurring of professional and legal boundaries. For example, the examiner is sometimes also a police officer or detective, making it difficult for subjects to discern whether the examiner is also one of the officers investigating them or an impartial outsider providing an objective view on the matter. The line is often intentionally blurred, so that examinees are at one moment led to believe that the examiner is impartial and might support their innocence or speak on their behalf to the police investigation team, and at other times made subject to further interrogation by the examiner in a more official police capacity. This is reflected in the way in which the lines between the polygraph exam and the interrogation are also blurred. The pre-test interview and post-test interview might smoothly transition into custodial interrogation of the suspect without a clear demarcation being made. The effect of this is that suspects might not realise they now require legal counsel to be present, since they have stopped being polygraphed and begun to be interrogated. This and other manipulations form a key part of the polygraph interrogation and how it is used to extract confessions. The techniques I outline are:

1 Keying objectivity;
2 Performing expertise;
3 Ontological and epistemic certainty;
4 Bleeding, clearing and composing.

1. Keying objectivity

The first technique, 'keying objectivity', draws materials and associated power relations from medicine into the polygraph interrogation to manipulate the subject into passivity and induce a confessional mode. The socio-historical analysis of lie detection, whilst powerful, relies predominantly on the accounts of the polygraph machine's inventors, who were much better characterised as scientists-cum-policemen than are contemporary polygraph operators, who tend to have significantly less scientific experience. Nonetheless, this research provides a useful argument regarding the use of a performance of 'scientist-doctor-investigator' and how this underlies the ability of lie detection examiners to elicit confessions (Balmer and Sandland, 2012; Littlefield, 2011; Bunn, 2012; Weber, 2008; Alder, 2007). To refine this argument in the context of interrogation transcripts, I read them in light of Goffman's various analyses of social performance and place more weight on the role of 'doctor-detective' than scientist-investigator. Emphasising the role-play of doctor-detective helps to better articulate how it is that the polygraph examiner is able to extract information and – in many cases – confessions from suspects. It also helps show why this performance has the potential to bring about false confessions. Goffman's (1974) notion of 'keying' proves essential to my framing of this literature, as I develop in due course. First, I review the important contributions made in the literature so far.

Alder (2002: 2) notes that during the early years of the polygraph's development, several of its proponents acknowledged that the lie detector would not 'work' unless its subjects believed it 'worked'. To convince subjects that the device works, its developers began to integrate elements of performance with the examination. They intentionally determined to stage the exam in such a way as to replicate the image of objectivity common to medical and scientific situations. Following Goffman's (1974) analyses of performance, we can understand this as part of the actors' attempts to 'frame' the lie detection situation so as to elicit confessions. As regards the polygraph examiner's role, for example, Bunn (2007: 174) argues that in order to perform scientific objectivity the 'preferred tactic was simply to appropriate stereotypical symbols of scientific activity: the white coat, the black box, the graph, the disembodied pursuit of truth.'

The research also points to the use of playing card magic, designed to trick the subject into thinking that the polygraph machine could guess the card that they had selected from a pack. This was rigged – for example by using marked cards or trick decks – so that the machine would always be right. In the face of the device's epistemological flaws, this magical performance was intended to convince the subjects that it was in fact infallible, when in reality it was not effective enough to rely on its own powers of lie detection in judging the subject's card.

Weber (2008: 74–75) draws attention to the spatial and material arrangement of the body, the examiner and the machine as being important in the performance of objectivity. She reports that subjects are generally made to face away from the examiner and the recording equipment so that they cannot see when and how they are being assessed. This positions the subject as an object under the gaze of the examiner, not a participant in a conversation or test. The Department of Defense Polygraph Institute (DODPI, renamed the Defense Academy for Credibility Assessment [DACA] in 2007), which trains government employees on university accredited courses in polygraph technique, recommends in its 1990s training manual 'Interview and Interrogation' that the polygraph interview space be arranged in very specific ways. This includes ensuring that there is no police equipment in the room to maintain a neutral appearance (implying that this is a medical, not a police, space) and that the exam take place in a private, small room (ensuring people feel isolated). Such arrangements of power are revealed clearly in the advice on chairs and tables, which suggests the suspect be given a chair lower than the interrogator's to limit him physically and psychologically, and that the table be small in width to heighten the suspect's awareness of the interrogator's proximity and elevation (DODPI, 1991: 32). All of these strategies enable examiners to combine performances of the roles of doctor and detective, constituting them as dispassionate, objective and unimpeachable witnesses to the truth of the subject's thoughts and feelings, as apparently recorded by the machine. It also frames them as powerful figures to obey.

The devices repurposed as lie detectors are just as important to the successful performance of objectivity as is the garb adopted by the examiner and the arrangement of the space. The fact that these technologies are borrowed from medical and scientific contexts means that they carry the weight of objectivity by virtue of affiliation with these other more familiar situations. It also evidences the importance of understanding how polygraph examiners play the role of doctor more than they do scientist.

A sphygmomanometer's rubber cuff wrapped around the subject's arm reminds the interviewee of being under the medical gaze, as do the various other physiological technologies measuring respiration and sweat. The charts, scores, algorithms and so forth all help to stage objectivity in ways which are familiar from hospital and GP visits, spaces common in the lives of everyday suspects. An example of this technique in operation can be seen in the quotation below, as the examiner explains the equipment whilst connecting it to the subject's body:

EXAMINER: [This is] a smaller version of the blood pressure cup that is used in doctors' offices. It's identical. It just makes it a little more comfortable for me to put it on your thumb.

SUBJECT. Okay.

EXAMINER: It accomplishes the same thing. These two black tubes over here, these are pneumographs. One is going on your upper chest. One's on your lower chest. . . . It measures minute reactions in your breathing caused by your autonomic nervous system when you are under stress. This device is called a plethysmograph. If you have ever been checked out in a hospital or a doctor's, they also call this a pulse oxymeter.

In Goffman's (1959) terminology, the polygraph examiner is 'keying' the medical situation, as part of a broader fabrication and 'containment' of the subject. The scientific language and the physiological devices assembled into the lie detection apparatus all represent 'props', borrowed from the frame of medicine, which frame the lie detection situation as having particular rules and people as having particular roles, much as they would in a hospital or doctor's office. The keying produces 'a systematic transformation . . . across materials already meaningful in accordance with a schema of interpretation' (Goffman, 1974: 45).

Keying the medical situation helps examiners to induce in those being interrogated the patient or subject role. This encourages suspects to be forthcoming, to talk, chatter about their actions, memories and so forth, and to do so to support the work of the investigator in objectively determining the truth of the matters at hand, much as they would if they were patients trying to assist in their diagnosis. However, unlike 'keys' of situations in which all participants are 'in' on the transformation (for example, when people 'play fight'), in the keying of the medical scenario for lie detection purposes, the emphasis is on 'containing' the subject, in other words, on keeping the subject in the dark about the transformation which has been conducted.

Furthermore, the props are used to link the subject's body to claims about truthfulness, becoming 'technological inscriptions' (Latour and Woolgar, 1986) that replace the body whilst speaking on its behalf. For example, in the quotation from a different interrogation below, the polygraph results chart comes to stand in for the truth of the subject's experience:

EXAMINER: I have had a chance to review your charts on both tests. That first test you did not pass the test. And it is pretty clear to me that you're not telling me the complete truth today.

These various strategies, choreographed through props and staging, are thereby used to frame the situation by keying medicine not simply to borrow the order of interaction but also the implicit power dynamic of expert and passive subject.

2. *Performing expertise*

In line with the doctor-detective role-play, examiners will generally give an account of their expertise as polygraph operators, providing a description of their experience (for example, number of tests conducted or years worked as an examiner) and outlining their qualifications. They often give an indication that they are or might be an expert in some area of science, be this psychology, psychophysiology or some such. These accounts of expertise help to affirm their status as objective agents determining the truth of the subjects' bodily responses. In the following example, the examiner draws on a range of different forms of experience and expertise to account for his objectivity and skill with the polygraph:

EXAMINER: As you know, I'm a licensed private investigator. I'm a retired state police lieutenant. I've been an investigator now for as long as you've been alive. Forty years. You're forty, right?

SUBJECT: Yes.

EXAMINER: I started my investigating career about the time you were born. . . . I've conducted thousands of examinations. As far as the polygraph, I'm a recent graduate of the New England Polygraph Institute, which is one of the four best schools in the world. . . . My class selected me as class president, and overall I graduated with an average of 96%.

. . .

So, putting all of that together the picture I'm trying to paint to you is I'm qualified to do this, and I'm really good—

SUBJECT: You know what you are doing.

The effect of providing this description of expertise confirms examiners' objectivity, but also their authority, ensuring a power dynamic in which subjects must do whatever they can to account for themselves, their past and feelings, their thoughts and actions, to 'pass the test'. Indeed, the framing of the examination as a 'test' in the first place, one administered by the examiner and taken by the subject, along with the idea that a subject can 'pass' or 'fail', exemplifies this obligation.

At the same time, the demonstration of expertise and experience has the effect of 'out-weighing' subjects' own claims to knowledge about their actions, past events and so on. The demonstration of polygraph operators' expertise and experience is used to show subjects how the examiners' future testimony would be framed if they were expert witnesses in court. It gives examinees an idea of what it would be like to go up against these people during a trial, people with, generally, much greater claim to authority on the truth, through their credentials, experience, professional ties and so forth. In this way, the exam becomes a 'trial run' to test the subject against the power of the examiner; failure now, it is implied, will mean failure later.

3. Ontological and epistemological certainty

By keying the medical descriptions of the body and performing medical tropes, the examiner works to convince the subject that the latter is unable to affect the results. The aim is to ensure that subjects do not believe that there is anything they can do to prevent the machine from recording their bodily responses accurately, and thereby determining whether they are being honest or deceptive. After giving a description of their expertise, examiners generally provide descriptions of how the body responds to telling a lie, and how the polygraph machine works, to achieve its lie detection function. This often invokes the so-called 'fight or flight' response, or relies on the distinction between voluntary and involuntary nerves, as in the following example:

EXAMINER: There are two sets of nerves in your body. . . . You're in control of your voluntary nerves. You can control those actions, but we also have another set of nerves, and this is where the polygraph comes into play. It's called the autonomic or the involuntary nervous system, and it's the one that works without your control.

SUBJECT: Okay.

EXAMINER: It's the one that keeps your heart beating, your eyes blinking, and you breathing. And I think the best example of the two, if I tell you to hold your breath, you can hold your breath. That's your voluntary nerves. But if [you] sit here and keep [your] mouth shut, after forty-five seconds you're going to start breathing one way or another.

SUBJECT: Yeah.

EXAMINER: It's your involuntary nerves overriding everything. You have no control over those whatsoever.

This account is typical. The examiner provides a description of how the subject has no control over his involuntary nervous system, often drawing on an example which might be familiar from everyday life, such as holding one's breath. The purpose is to convince the examinee that trying to beat the polygraph is as futile as trying to hold one's breath indefinitely: you can't keep schtum forever. In another case the examiner describes the experience of playing sports, recounting how the body responds automatically to exercise and linking this to how it automatically responds to telling fibs. By connecting everyday experiences of bodily ontologies to lying and lie detection, the subject is led to believe in the certainty of a particular ontological account: a causal chain is drawn from lying to breathing and sweating, as if the heartbeat thumped out the rhythm of truth regardless of the mind's designs.

Related to the description of how lies and the body are connected is a description of how the polygraph machine works to detect these lies, namely by objectively recording the body's responses:

EXAMINER: What I want you to keep in mind, this polygraph instrument is just a recording unit. All it can do is record your physiological responses, and, if you leave out any part of the truth, and tell me the partial truth, because of your autonomic, involuntary nervous system, it's going to react significantly, and I'll know you're not telling me the truth. It's really as simple as that. The key is knowing and understanding the polygraph has no human capabilities. It is simply an impartial, scientific instrument.

In the quotation above, the machine is denied any agency in the polygraph examination process, for it is 'simply an impartial, scientific instrument'. The representation of the machine's passive recording of physiological phenomena is used to instruct subjects not only in the objectivity of the process, but also in their continued passivity. If the machine has no human qualities it cannot be reasoned with, manipulated or betrayed. The irony of its revelation of truth is that the machine's perfect capacity (to be able to see into the other's intentions) depends upon its apparent lack of agency. That it can be used to accurately describe the subject's emotions depends on its lack of feeling. It sees lies because it cannot lie. At least, that is what subjects are instructed. There is, however, a deep ontological uncertainty in polygraph science literature. There is very little evidence that the body responds in a reliable fashion to telling a lie, and the scientific as well as the legal literature continue to debate whether the machine can, therefore, ever tell whether it is measuring a lie or something else entirely. All such uncertainties are erased from the accounts of the machine during the examination. Although these descriptions are provided ostensibly to explain how the machine will work during the test, they are rather more geared towards making the subject fear the device. In the pre-test and post-test interview phases, these ontological and epistemological descriptions will be used to try to elicit talk and extract confessions.

4. Bleeding, clearing and composing

Three related techniques form the backbone of the confessional, interrogational form of interaction which polygraph examinations take. I term these processes 'bleeding', 'clearing' and 'composing'. Bleeding refers to the examiner's use of certain techniques to elicit an account of the examinee's memories, feelings and fears. Clearing refers to the use of talk elicited in bleeding to anticipate and thereby close down the ways in which subjects might try to contest the results of the examination. Composing refers to the way in which the examiner uses the test and the interrogation to frame and guide the talk produced by the examinee into accounts of people (including the subject), actions and events.

The processes are not linear, although an examination tends to begin with a progression of bleeding, clearing and composing in the pre-test phase. The real power of these techniques is in how they are skillfully connected by the examiner, moving between them to manipulate the subject into speaking and confessing in the pre-test and post-test interrogations. Bleeding and clearing are sometimes explicitly presented as parts of the exam, with some examiners using these terms in their explanations. At other times, they are conducted without being named. I adapt the situated use of the terms bleeding and clearing to my own purposes. Composing is the hidden aim of the interrogation and is not referred to in the usual course of an examiner's affairs.

Bleeding

First, subjects are encouraged to 'bleed off' their emotions during the pre-test interview. After examiners tell subjects about how the polygraph works, providing ontological and epistemological accounts of lie detection, they proceed to instruct subjects in what they must do to successfully take and pass the test. The example below is indicative:

EXAMINER: I'm going to explain what you have to do to have a successful polygraph examination this morning, and there are three keys you need to be aware of, and I will be repeating them. One is you have to be one hundred percent honest with me about every question I ask. And one hundred percent means one hundred, not ninety-nine point ninety-nine. . . . The polygraph cannot tell the difference between most of the truth and none of the truth . . . because of the physiological reactions of your body . . . you cannot withhold any information.

As in this quotation, subjects are told that they must be completely honest to pass the test because the machine cannot tell a complete lie from a partial truth. Its epistemological failure in this regard is used to examiners' advantage rather than their detriment. The description of the polygraph's mechanics and limitations is turned into a powerful tool for leveraging information by constructing a power relation which aligns truth and risk in a specific way.

People taking a test are told that they cannot risk concealing any information which they can think of which might be related to the questions they will be asked during the exam, or which might have any bearing whatsoever on the case at hand, because the polygraph will objectively demonstrate that they are concealing something, but will not be able to judge whether this concealed information is significant to the question or not. In other words, any concealment is constructed as a risk which the subject is taking, under threat of receiving a 'deception indicated' result on the test and thereby 'failing' the exam. Risk is linked to truth in such a way that it is not only specific truths (yes or no on the test questions) which need to be told, but rather total honesty which must be performed. Subjects are made to feel that to pass the test they must confess themselves in as much detail as possible.

With the risk-truth power relation in place, the subject is taken through a process of 'bleeding off' their emotions and thoughts, which, they are told, is designed to help them pass the test by 'getting off their chests' anything that they feel guilty about (past or present), anything they have lied about so far in the investigation, and anything about which they might not have been completely forthcoming. As they talk through these issues in the pre-test phase, subjects are encouraged to believe that they have 'let go' of those thoughts and feelings and that even if they come up during the test phase, they will no longer interfere with the results. The following is an example:

EXAMINER: There are questions here we will get to later called safety valve questions. They're designed to bleed off that emotion.

SUBJECT: Okay.

EXAMINER: So when I ask you what you did or did not do to [the victim], I don't want something else going through your mind.

SUBJECT: Okay.

EXAMINER: . . . By the time we get to the in-test phase when I'm asking you the questions there should be no unknowns in your mind.

Bleeding is thereby conducted (explicitly, in this example) as a way of helping examinees to pass the test and demonstrate their innocence, but its most immediate and hidden effect is to help the investigation by eliciting more talk under the threat of failure. It is in this way that the pre-test interview forms a crucial part of the polygraph's confessional form, for the description of how the polygraph machine works, regarding its combination of powers and shortcomings, is used to create a risk-truth power relation which compels subjects to tell as much as they can think of. Before the machine is even switched on, the manipulation has begun.

Examiners deliver the bleeding process differently. Some provide examples of the kinds of issues which might affect the test, for example by listing broadly irrelevant issues that might confuse the machine: 'you might have lied as a child'; 'you might have seen something you shouldn't have at work'. Others invent hypothetical secrets which subjects might be keeping or lies they might have told in direct relation to the issue under investigation: 'you did not do this, but you do know who did'.

Such tactics are designed to elicit further talk, to invite subjects to answer questions and provide information which they might not otherwise answer or supply, both in general terms regarding the subject's life and in specific terms regarding the issues at hand. This is not always about information which is incriminating to subjects, for it can be just as likely that the police use a polygraph interrogation to compel witnesses known or suspected to be innocent to provide information which could be of use in removing them from suspicion or in pursuing (and later prosecuting) another suspect.

Clearing

Subjects are given to understand that the bleeding-off process will result in their 'clearing' their minds of any unwanted thoughts that might interfere with the test. Having disclosed anything they can think of, they are told that this information will no longer cloud the issue under investigation. This is framed as an important step in passing the polygraph test and one which requires total disclosure:

EXAMINER: If you haven't been a hundred percent truthful with the police or me—

SUBJECT: Right.

EXAMINER: you won't pass this examination. . . . You need to share with me the complete truth so we can resolve the matter. I need to make sure I get the questions—

SUBJECT: Right.

EXAMINER: Right. That's the critical thing, and as long as you are telling me the absolute unadulterated, unedited one hundred percent truth, you'll have no problem.

Clearing is the explicit aim of bleeding, and subjects will be moved into the test phase once the examiner says they are clear. Strategically, however, getting the subject to a 'clear' state does not aim at improving the accuracy of the questions and the test procedure so much as it does at making it difficult for the subject to contest the results later.

Clearing is used to anticipate and nullify subjects' possible justifications, excuses or explanations for why they might receive a 'deception indicated' result on test questions. This allows examiners to force subjects' hands in the post-test interview to provide even more information (assuming they are innocent) or to try to extract a confession (assuming they are guilty). In the post-test phase, examiners can argue that subjects were clear when they took the test, and so a deception indicated or inconclusive result is now framed as being a product of concealment rather than a limitation of the polygraph. Clearing thereby supports the bleeding technique by requiring further talk to explain the results. It might be that examiners conduct multiple tests on subjects, each time further bleeding them to clear any memories, feelings or thoughts which subjects think might have interfered with the previous test, gradually removing all possible contest of the final test results.

Composing

The talk elicited in bleeding and clearing is then used in the process of 'composing',[4] by which I mean the construction of accounts of people, actions or events which are relevant to the case. Such accounts are produced by the examiner, using the words of the subject, but also by reworking them into descriptions which serve the examiner's ends. They are given back to the subject in the talk of the examiner to provoke the suspect into agreeing with the accounts or disagreeing with them, and by doing so, into providing more information to try to influence the examiner to rework the account to be given back to the police.

The process of composing is especially powerful in-between test phases, or in the post-examination interview, when subjects are often told, irrespective of the results, that the test indicates they are being deceptive or not completely truthful. Sometimes software is used to automatically score the chart, providing an immediate outcome which can be shared with the subject (e.g. if it is 'no deception indicated', 'inconclusive' or 'deception indicated'). Sometimes examiners will provide a quick outcome by rapidly scoring the chart based on their own technique. And sometimes examiners will simply say that there is deception indicated in the absence of any formal measurement of the charts.

The point of providing subjects with a deception indicated outcome (whether from the results or by fabrication) is to encourage them to continue to provide information which can be used to compose accounts of people, actions and events to share with the police. Compared with pre-test or intermittent bleeding, clearing and composition, in the post-test phase the results from the test and the mythic qualities of the polygraph machine can be used powerfully in a more directed and aggressive interrogation. For example, in the following quotation from an interrogation in which the examiner has just told the subject that he failed the test:

EXAMINER: I'm laying out for you what I've seen here. You can't argue with this. I've done over 2,000 of these. When I told you I know what I'm doing, I know what I'm doing.

SUBJECT: Well you must have slipped up today, I mean I don't know.

EXAMINER: What do you think, the chances are better that you slipped up and the truth snuck out here, or that I slipped up?

SUBJECT: I know you slipped up. But if you're telling me that I [pointing to the machine at his side] failed that test—

EXAMINER: I can tell you in no uncertain terms that you failed that test.
SUBJECT: Well, it wasn't valid.
EXAMINER: Absolutely valid.

The post-test phase can include such attempts by the subject to contest the examiner's compositions. However, the power relations established through keying the medical situation are brought together with the bleeding, clearing and composing processes, making it difficult for the subject to sustain his resistance. Most immediately, the examiner presents the results as a fact indicating a specific composition of the crime, one which the subject must account for through further talk.

Combining the techniques

A good examiner will skillfully, but deceptively and manipulatively, weave together the techniques of bleeding and clearing with the performances of objectivity and expertise, draw in accounts of ontological and epistemological certainty and just as often directly involve the machine or the charts to compose an account of the subject's deceptiveness, all of which is done to induce further talk. The example below evidences all of the techniques brought together:

EXAMINER: You know when someone fails a polygraph [he reaches for the laptop] . . . That is your heart and your mind betraying what your body knows to be the truth. My masters degree is in forensic psychophysiology, which means the mind and the body interacting, to determine truth in an interview setting. That's what I do. This [pointing to the chart on the laptop screen] is actually admissible in court because it is scientifically valid. So when I say you know more than what you told me, I'm not guessing that.

In this fashion, the post-test phase is used in relation to the work done in the pre-test phase. The examiner leans on the idea that the subject was cleared during the pre-test phase, and so can challenge the examinee to explain why the results suggest deception. Indeed, the examiner will often allege that the subject is lying about something covered in the test and that this is now known and needs explaining, without indicating on which questions the subject is supposed to have lied. For example, in the following exchange, the examiner tells the subject that the result is 'deception indicated' and presses him to explain why he is lying, as the examinee tries to keep some claim over his account in the face of the examiner's broad compositions:

SUBJECT: Does the test show, what does, I mean you're talking crazy . . . the accusations you're making.
EXAMINER: Yeah, I'm saying you were involved in this.
SUBJECT: Absolutely not. No.
EXAMINER: Well, you've already told me that. That's why we did the polygraph to find out. Well, now we both know.

SUBJECT: I've told you what now?
EXAMINER: Well, you've told me before you had nothing to do with it.
SUBJECT: Yes.
EXAMINER: Well, now we're passed that. Because I've already shown that you did have something to do with it.

. . .

SUBJECT: But I'm telling you I didn't have—
EXAMINER: And I'm telling you you're lying.
SUBJECT: I—
EXAMINER: I can't make it any clearer. I'm not trying to hurt your feelings.
SUBJECT: I—
EXAMINER: I'm telling you the truth.
SUBJECT: That's what I'm doing to you, I'm telling you the truth.
EXAMINER: No, you're not. We're passed that. I've already shown you.

In the quotation above, the subject tries to contest the examiner's compositions about the clearing process and the results of the test, maintaining his veracity as the examiner continues to undermine this account, attempting to provoke a confession. Interrupting the subject signifies that any resistance to the composition in which the subject is described as being deceptive will not be taken seriously. Eventually, subjects must explain why they are being deceptive (or why the machine might say they are being deception) if they wish to be believed, or they will be considered to have failed the test.

Although it is represented as an objective device which pinpoints lies through 'yes' or 'no' answers to clearly worded questions, the real power of the polygraph lies in its epistemological failures. The body is not used to speak the truth. Rather, the polygraph interrogation plays on the uncertainties in the connections between lying, body and truth, leveraging the ambiguity of 'deception indicated' under threat of failure, all to elicit talk. On the face of it, such extraction of information through use of deception, threat and manipulation might seem a fair method for getting confessions from guilty subjects. However, the relations of power established in the production of talk, when applied to subjects in vulnerable situations, being sleep deprived, hungry and in withdrawal, can prove a heady mix of emotions which do not necessarily induce truthful talk. Rather, they can prompt subjects to make false confessions.

In addition, innocent subjects might confess to a more minor crime fearing that they are – having apparently failed a test – going to be found guilty of a more serious crime (Leo, 2008). More troubling still, vulnerable and innocent subjects might come to believe, even if only for a moment, that they actually did commit a crime. Once a verbal confession is made, it is practically impossible to retract and is used in powerful ways to secure convictions during trial.

The case of Peter Reilly proves a tragic example. In the final section of this chapter, I outline how the tricks and manipulations inherent to lie detection interrogation practices were used to induce a false confession of murder from Peter, an eighteen-year-old resident of Connecticut in the 1970s.

The interrogation of Peter Reilly

Studies show that people find it difficult to accept that others could come to believe they were guilty of and confess to a crime they have not committed (Kassin and Wrightsman, 1980). However, there is ample evidence that innocent people do sometimes confess to crimes having come to believe that they were the guilty party (Rattner, 1988). Peter Reilly's case exemplifies this problem.

At around 7.30pm on Friday, 28 September 1973, Reilly, a gangly eighteen-year-old, left home saying goodbye to his mother, Barbara, who sat watching the TV. Peter was off to a meeting at the local Methodist Church in Canaan, Connecticut. He drove home around 9.50pm, after dropping off a friend on the way. After opening the front door he called out for the whereabouts of his mother but received no reply. He turned right from the front door to find Barbara's body, severely mutilated and almost naked on the bedroom floor. After a lengthy interrogation he confessed to the crime and was found guilty at a subsequent jury trial.

Peter's case and his life after the trial are described in significant detail in Connery's (1977) engaging book, *Guilty Until Proven Innocent*. It is an apt title, for Reilly was treated as the only suspect in his mother's murder, though there was no material evidence tying him to the crime. First, although Barbara's body and the surrounding area were covered in blood (she had been nearly decapitated, had both her legs broken, and been soaked in water) there was no blood on Peter's clothes whatsoever, which he had been wearing since earlier in the day. The timeline also made it hard to see how he could have committed such an elaborate murder. His mother had called her doctor only fifteen minutes or so before Peter got home, and in his appeal a forensic pathologist would testify that there was not enough time for someone to have committed this murder in this way and then clean himself thoroughly, arrange for an ambulance and contact other people on the telephone. Finally, there was no motive for him to have committed the crime in such a spontaneous and brutal manner. In addition, there was evidence that someone else could have been in the house and committed the crime. Peter reported that the front door was open when he arrived home and the police found the backdoor open as well.

It became a sensational case, reported on in national newspapers and, once Peter was freed, eventually led to calls for an investigation into policing practices in Connecticut. The jury's guilty verdict seemed to have been based predominantly on Peter's signed confession, but the interrogation by the police was deeply deceptive, with a great deal of manipulation and subterfuge, conducted over a long enough period to exhaust any boy eighteen years of age, and with the aid of the polygraph machine, applied in a most egregious fashion. Peter came to believe that he had murdered his mother in spite of his own doubts, in spite of the evidence and in spite of the results of the polygraph test (which had not been properly analysed when they were presented to him and thus could not indicate deception). He provided a false confession which put him in prison at the most vulnerable moment of his life. In this final section of the chapter, I explore Peter's interrogation, drawing on the analysis in the preceding section.

There are, broadly, three kinds of false confession (Kassin and Wrightsman, 1985): 'voluntary', meaning a confession given willingly and without coercion; 'coerced compliant' confessions consciously designed by the subject to avoid further interrogation or harm, or to access promised benefits (such as a reduced sentence); and finally, 'coerced-internalised' in which the subject comes to believe that he is guilty of the crime. Peter's false confession was of the third kind. It is such cases in which, I propose, the polygraph gains its greatest power (of psychological manipulation) and its greatest threat (of convincing innocent people that they are guilty).

Existing studies in psychology and law have used experimental techniques, mostly in low risk situations, to explore the processes through which people can be led to inculcate false memories. Early studies, for example, showed that the wording of interview questions could affect how subjects remembered events that they had witnessed (Loftus et al., 1975). Through the subsequent provision of misinformation about events they had directly observed, subjects could be induced to produce accounts of false memories which reflected the misinformation (Loftus and Hoffman, 1989). These false memories were also shown to be just as powerful if not more powerful than the original memories of the events in question (Loftus et al., 1989).

Over several decades, important social psychology research helped to demonstrate that our own memories of events can be altered or added to depending on how we are asked to recall them and especially when we are given misinformation about what happened and asked to confirm or deny such accounts. It influenced legal scholarship and rulings, by troubling the reliance on eyewitness testimony, but it has had far less success in undermining the power of confessions induced through questionable, manipulative and deceptive means. This remains a problem in the US legal system, for such police interrogations are common and can produce false memories which lead to wrongful convictions.

For example, in an influential study, Kassin and Kiechel (1996) demonstrated the importance of compliance, internalisation and confabulation in the production of false confessions based on false memory. Compliance represents the subject's agreement (e.g. by signing a confession) with the false account produced by an authority figure. Internalisation refers to the subject's adoption of the false account as a memory and the acceptance of associated guilt of having committed the actions described. Confabulation is the process of reproducing the false account as one's own memory of events to demonstrate one's guilt and explain how and why it was that one committed the act.

Outlining Peter's case, I argue that the techniques of 'performing the doctor-detective', 'demonstrating expertise', 'ontological and epistemological certainty' and 'bleeding, clearing and composing' each have the potential to contribute to the production of false confessions.

The examiner who conducted Peter's polygraph interrogation adopted the performance of doctor-detective, positioning himself as non-partisan and keying the medical situation. For example, as he connected the blood pressure cuff to Peter's arm, he explained:

> This comes from a medical supply house. It's the exact same thing as what a doctor uses. You see, this little gauge goes right here, . . . We're measuring your blood pressure, your heartbeat, your pulse, you see, because that's very important because that's the only muscle in your body you can't control.

By keying the medical situation, the examiner attempts to establish a trusting relationship with the subject, but one which mirrors the doctor-patient power structure, encouraging the subject to speak and to take as objective fact the examiner's findings during the test.

This poses a risk to vulnerable, suggestible subjects, for it can lead them to take the examiner's statements as dispassionately truthful, when in reality they are often loaded, leading and deceptive. Before the formal test begins, the examiner repeatedly instructs Peter that he will not lie to him, deceive him or trick him, and that he is here only to find out the truth of what happened using the polygraph as an objective test. However, the interrogation is rife with misleading information and outright lies. For example, prior to the test phase, the examiner uses the 'numbered card test' to convince Peter of the objectivity, validity and reliability of the polygraph machine. The examiner shows him what appears to be a shuffled deck of numbered cards, allowing him to select one and then take the pack to see that they are not marked or dubious. However, the numbered card test depends on a different deception, 'sleight of hand', meaning the examiner switches the shuffled deck for a pre-arranged deck before allowing Peter to take one. As such, the examiner knows exactly which card Peter has selected before beginning the polygraph test, so that when Peter is shown the results, all of his bodily responses represented in the chart can be presented to him as evidence that the machine could read his mind and tell which card he selected. This is all a fabrication based on deceptively scoring the chart to reflect the known card selection. The numbered card test is used to convince Peter that he is a 'textbook subject', meaning that the test will work effectively and validly, and that he cannot beat it:

EXAMINER: You're a textbook reactor. When you tell a lie you go right to the top of my chart. . . . You're perfect. In other words, you are a good reactor.

Moreover, the examiner deceives Peter regarding the scientific status of the lie detector by using statements of ontological and epistemological certainty. In the following example, the examiner misleads Peter as to how the machine works:

EXAMINER: That's why I have this [the polygraph machine]. That reads your brain for me.
SUBJECT: Does that actually read my brain?
EXAMINER: Oh, definitely, definitely. And if you've told me the truth this is what your brain is going to tell me.

Peter is highly compliant, providing as much information as he can, going over his account of events repeatedly and consistently, answering questions in a manner which anticipates help with finding the murderer of his mother. The examiner uses this compliance to deliver the bleeding and clearing process, taking Peter through a series of question-preparation exchanges, either by asking Peter to explain his account or by providing him with information about the crime scene. These exchanges are used to get Peter to talk and to get him to confirm or deny information in his account or knowledge of the scene, under the guise of producing the exact wording of the questions. For example, during the pre-test phase the examiner prepares a question about Barbara's body using information provided by the autopsy, of which Peter is unaware. He is given graphic and puzzling information about his mother's murder.

EXAMINER: Now, she had two broken legs. All right?
SUBJECT: She did?
EXAMINER: Right. Just above the knee. Okay? Now, this had to happen some way. Right?
SUBJECT: Right.
EXAMINER: Do you have any feelings about this?
SUBJECT: I don't know, I didn't know, I didn't know what it was.

Each question Peter is asked is used to bleed further talk from him, and then to clear him on each of these issues. The examiner also uses a broad bleeding and clearing question to use in the post-test phase. He asks Peter whether his statement given to the police the night prior was the truth or a lie, getting a confirmation to prepare for the test question and then clearing him on this broad issue:

EXAMINER: And again, you gave a statement to the police last night, right? So, if I ask you that question on the polygraph: is the statement you made the truth?
SUBJECT: Yes.
EXAMINER: Okay. Again, when I say that, I don't want you sitting here, thinking, well, gee, was it nine-thirty-five I got home or was it nine-thirty-six I got home. I'm not interested in that. I mean the overall statement that you gave as to what you had done about arriving home and finding your mother on the floor.

Peter continues to comply with the question formation process, giving as much information as he can. But he is also nervous, and often tries to clarify how the test works to try to make sure that he will not erroneously be seen to be withholding or deceptive. His worries are exacerbated when the examiner spooks him by deploying the 'epistemological failure' strategy of bleeding:

EXAMINER: Now, like I said, 95% of people that come in here, when they lie to me, they give me a big reaction over here [indicating the polygraph chart]. Okay? Because this is a machine, it isn't a human being. This can't tell the difference between whether you stole ten dollars or whether you stole a million dollars.

Peter fears that the machine will misread his mind and expresses some concern as the examiner begins to put together guilty compositions in the pre-test phase. For example, the polygraph operator hypothesises that Peter might have been acting in self-defence:

EXAMINER: Now maybe you and your mother had an argument and one thing led to another. She attacked you, I don't know. Won't be sure until we finish. If anything like that happened I'll know it over here on these charts.

Nonetheless, Peter continues to supply information, trying to assist the doctor-detective to get evidence to support his innocence and to avoid failing the test.

The manipulations become more dangerous as the examiner concludes the test phase by slipping in an additional, unprepared question at the end of the third test, to use Peter's response to this surprise question to claim that he is concealing information and lying about hurting his mother:

EXAMINER: Do you have a clear recollection of what happened last night?
SUBJECT: Yes.
EXAMINER: Is there any doubt in your mind, Pete?
SUBJECT: Can you stop the test?
EXAMINER: Okay.
SUBJECT: I didn't understand that last question. I didn't understand—
EXAMINER: I think we've got a little problem here, Peter.

In the post-test phase, the examiner lies to Peter, saying that the charts demonstrate he is being deceptive or concealing information about the events of his mother's murder, but the test had not been properly scored and the exam used a trap (an unprepared question), rendering it void. It is in the post-test phase that the examiner draws together the bleeding, clearing and composition processes, challenging Peter to explain the apparently incriminating charts, given he was cleared on the questions:

SUBJECT: What about the question that says did I speak to my mother?
EXAMINER: That's bothering you too.
SUBJECT: It is?

EXAMINER: You whipped out on that question.

SUBJECT: Let me see, what about the— You know what bothers me on that one, though, is because of my yelling to my mom [to see if she was home as I entered the door]. When you ask me that question that's the first thing I think of.

EXAMINER: Yes. But I explained to you what I meant by that.

SUBJECT: Right.

EXAMINER: Remember, we talked it over and I said did you have con— when I said [we mean did you] have a conversation, and you said 'yes' I understand what you mean by that.

As in the example above, the examiner continues to put forward compositions, using Peter's words, evidence from the crime scene, his own fabrications and hypotheses to induce further talk from Peter. Compositions the examiner puts forward to Peter include the following contradictory and unfounded statements:

- You did it and you know you did it, but you are too ashamed to say so. That is why the charts say you are lying.
- You did it and you are in denial, but the polygraph can read your hidden thoughts. It shows that you did it.
- You did it and you are too scared to say because you fear the consequences. But you want to tell me and that is why you came to have this test, which you knew would show you did it, so that you could confess.
- You did it but cannot remember. You flipped out and now there is a gap in your memory. But the polygraph shows you have the memories still and you can access them if you want to.
- You accidentally hit your mum with the car and tried to cover it up by making it look like a murder. That is why the polygraph is showing that you are concealing information.

The result of these various compositions is a great deal of epistemological mess. The examiner's compositions of how Peter might have committed the acts and why the polygraph charts show he is being deceptive cannot be reconciled with each other. They also demonstrate how epistemological and ontological uncertainty work results in powerful ambiguities when conducted in the context of interrogations.

Combined with the clearing process, which nullifies Peter's attempts to account for the polygraph's 'deception indicated' results by suggesting that the machine has not worked properly, this epistemological mess and uncertainty work results in a set of competing ambiguities which the examiner can use to close off any possible attack on the objectivity of the results or of the test procedure. The effect on Peter, however, is tragic, for in trusting the doctor-detective examiner he has no recourse to explain the results except to accept the composition of a 'gap in your memory' because only this allows him to reconcile the claims made on his mind by the polygraph machine with his own recollections.

The examiner leverages epistemological certainty against Peter's recollections and against his appeals for other forms of evidence:

SUBJECT: But, if I did it. I mean, if I did it and I didn't realise it, there's got to be some clue there in the house anyway—
EXAMINER: Yes. Right.
SUBJECT: —that's gonna connect with it.
EXAMINER: But that's what I'm saying. I've got this clue here.
SUBJECT: Right.
EXAMINER: See [indicating the chart], this is recording your mind.

Once Peter begins to show some acceptance of this possibility, the examiner links this composition to his performance of expertise and enrols the polygraph charts into his explanation to push it further:

EXAMINER: As I say, I've met a hell of a lot of people in my life. Now, if you did it, it was a split-second thing that you did. You lost your head. . . . These charts say you hurt your mother last night.
SUBJECT: But the thing is, I don't remember it.
EXAMINER: The charts don't say that, Pete.

He then links the 'gap in your memory' composition to the technique of minimisation common to the Reid interrogation process. The examiner implies that since Peter did it accidentally or involuntarily, the punishment will be lighter, and explicitly states at one time or another that Peter will be out of prison in a few months, that people will understand why he did it, and that he will get medical help instead of imprisonment.

The effect of these combined techniques starts to influence Peter's accounts of his actions, beginning the process of internalisation. It is clear, for example, as Peter continues to provide information about his past relationship with his mother (explaining prior arguments and fights, how she was controlling, how he wanted to move out and wished he had gone to college), that he begins to connect the guilt he feels for these earlier memories to the guilt being attached to him by the examiner. He begins to doubt his own memory and to adopt the 'gap in the memory' composition:

SUBJECT: Would it definitely be me? Could it have been someone else?
EXAMINER: I don't think so, Peter.
SUBJECT: No.
EXAMINER: From what I'm seeing here. No way, from those reactions.
SUBJECT: Now I'm afraid because I was so sure, you know, that I didn't do something.

Once accepting the possibility, Peter, as with the experimental subjects in social psychological research, starts to parrot back the authority's composition, internalising and confabulating the false account of the murder of his mother. A particular exchange, in the midst of repeatedly going over Peter's account of the events, proves instrumental. The examiner seizes on something Reilly has been saying all along, to add to the composition. Peter says that as he entered the room he saw the bed where his mother had been reading, and for a moment thought that he saw her there in the bed with her book, but then he did a double take, and realised she was lying on the floor to the side of him instead.

SUBJECT: Then I took a double take—
EXAMINER: See that? Now that—
SUBJECT: That's something that could mess me up.
EXAMINER: Right here! See?
SUBJECT: Where I thought I took— made the double take.
EXAMINER: Right. See? I think this is probably where you flipped over a little bit.

The examiner tells him that the polygraph confirms the memories are there and that he flipped out and killed his mother. Peter begins to parrot the account, all the while still trying out possible explanations for how it could be that he does not remember committing the murder when the polygraph says that he did. But these are all nullified by clearing and ontological certainty, so that Peter begins to accept the idea that he has murdered his mother in a fit of rage and that he must now try to remember doing it in order to start recovering and remain psychologically healthy. Instead of continuing to explain how the polygraph could be wrong, since this only leads to techniques of nullification, Peter begins asking for help in getting the memories to come up. He asks if truth serum could help, if further polygraph tests would reveal the memories, or if a psychiatrist could get them out of him, because he cannot find any memories in his head of hurting his mother. The examiner tells him fatalistic stories, that Peter cannot help his mother any longer, that she cannot be brought back, that he will eventually remember what he did and that if he is going to have any kind of life in the future he must get the memories out today. Repeatedly closing Peter's attempts to challenge the results eventually works, and Peter states 'I believe I did it now.' He is once again taken through the examiner's composition, step by step, to prepare him to reproduce it in his own words. Peter even adds information to the examiner's composition, filling in glaring gaps in the logic and evidence. Once he is effectively confabulating the false memory, the examiner takes him back in to see the police officers and Peter signs his false confession.

Conclusion

One of the most intractable consequences of barring polygraph evidence from trial is that it has also barred much of the discussion of how polygraph-induced confessions were produced, expunging some of the deceptive and coercive practices of police interrogation from those cases in which exposure of these issues would be most crucial. The polygraphs' censure has the concurrent and damaging effect of hiding the more deceptive parts of the technique from the jury's critical gaze. This often has detrimental effects, for it excludes the apprehension of one of the most powerful and dangerous outcomes of polygraph interrogations when they are combined with practices of police deception: the interrogator can convince the innocent that they are guilty.

The most paradoxical issue in the heart of polygraph examinations, amidst a range of ironies and doubled meanings, is that the polygraph interrogation's ability to bring about a false confession demonstrates that the fundamental ontological conceit of lie detection is itself false. The machine does not simply 'read the mind', for it can be used to change it. Our memories, thoughts and feelings of guilt, our very beliefs about ourselves and our actions are malleable. They are not set into the stone of an unchanging past or firmly founded on an unmovable base of truth which can be read without re-writing. When police officers and operators lie about the mechanisms and results of lie detector tests, when they assert such a powerful claim over the truth of someone's body and thereby claim access to their thoughts, it is not without effect upon the person's mind. The lie detection examination can change our perception of ourselves and of our pasts, and thus it demonstrates its own false premise. This is especially the case when subjects are vulnerable, both physically and mentally, to suggestion or to self-compromise.

Scientific accounts and media representations of lie detection invoke an objective device, used to probe the body without brutality when guilty people will not confess. But it is not scientific certainty which has replaced the third degree. The reality is that the polygraph's power lies in the manipulation of uncertainties. The involuntary, false confession has become the tragic mirror of the involuntary nervous response.

Notes

1 *The People v. Charles Edward Storm*, 28 Cal. 4th 1007; 52 P. 3d 52, 58 (2002).
2 Ibid., 58.
3 *State v. John Kolander*, 236 Minn. 209, 52 N.W.2d 458 (1952).
4 In this analysis, I prefer the term 'composition' to the perhaps more obvious term to use, 'construction'. Composition helps to avoid a link to constructionism, which is not intended here. Compared to construction, composition also does a little more to highlight: (1) the way in which the compositions are assembled from a range of ingredients; and (2) the movement and shifting nature of the compositions, which are handed over and returned, changed and adjusted as the situation develops.

References

Alder, K. (2002) A Social History of Untruth: Lie Detection and Trust in Twentieth-Century America, *Representations* 80 (1): 1–33.

Alder, K. (2007) *The Lie Detectors: The History of an American Obsession*, New York, NY: Free Press.

Balmer, A.S. and Sandland, R. (2012) Making Monsters: The Polygraph, the Plethysmograph and Other Practices for the Performance of Abnormal Sexuality, *Journal of Law and Society* 39 (4): 593–615.

Bryan, I. (1997) *Interrogation and Confession: A Study of Progress, Process and Practice*, Aldershot: Ashgate Publishing Limited.

Bunn, G.C. (2007) Spectacular Science: The Lie Detector's Ambivalent Powers, *History of Psychology* 10 (2): 156–178.

Bunn, G.C. (2012) *The Truth Machine: A Social History of the Lie Detector*, Baltimore, MD: Johns Hopkins University Press.

Connery, D.S. (1977) *Guilty Until Proven Innocent*, New York, NY: Putnam.

DODPI (1991) *Interview and Interrogation*, Washington, DC: Department of Defense Polygraph Institute.

Goffman, E. (1959) *The Presentation of Self in Everyday Life*, Edinburgh: University of Edinburgh.

Goffman, E. (1974) *Frame Analysis: An Essay on the Organization of Experience*, Boston, MA: Harvard University Press.

Hunt, J. and Manning, P.K. (1991) The Social Context of Police Lying, *Symbolic Interaction* 14 (1): 51–70.

Inbau, F.E. and Reid, J.E. (1964) The Lie-Detector Technique: A Reliable and Valuable Investigative Aid, *American Bar Association Journal* 5 (5): 470–473.

Inbau, F.E., Reid, J.E., Buckley, J.P. and Jayne, B.C. (2013) *Essentials of the Reid Technique: Criminal Interrogation and Confessions* (2nd Edition), Burlington, MA: Jones & Bartlett Publishers.

Irving, B. and Hilgendorf, L. (1980) *Police Interrogation: A Case Study of Current Practice*, New York, NY: Crown Publishing.

Kassin, S.M., Appleby, S.C. and Perillo, J.T. (2010) Interviewing Suspects: Practice, Science, and Future Directions, *Legal and Criminological Psychology* 15 (1): 39–55.

Kassin, S.M. and Kiechel, K.L. (1996) The Social Psychology of False Confessions: Compliance, Internalization, and Confabulation, *Psychological Science* 7 (3): 125–128.

Kassin, S.M. and Wrightsman, L.S. (1980) Prior Confessions and Mock Juror Verdicts, *Journal of Applied Social Psychology* 10 (2): 133–146.

Kassin, S.M. and Wrightsman, L.S. (1985) Confession Evidence, pp. 67–94 in S.M. Kassin and L.S. Wrightsman (eds.) *The Psychology of Evidence and Trial Procedure*, Beverly Hills, CA: Sage Publications.

Larson, J.A. (1925) Present Police and Legal Methods for the Determination of the Innocence or Guilt of the Suspect, *Journal of the American Institute of Criminal Law and Criminology* 16 (2): 219–271.

Lassiter, G.D. (2006) *Interrogations, Confessions, and Entrapment*, New York, NY: Springer Publishing.

Latour, B. and Woolgar, S. (1986) *Laboratory Life: The Construction of Scientific Facts*, Princeton, NJ: Princeton University Press.

Leo, R.A. (2008) *Police Interrogation and American Justice*, Boston, MA: Harvard University Press.

Littlefield, M.M. (2011) *The Lying Brain: Lie Detection in Science and Science Fiction*, Ann Arbor: University of Michigan Press.

Loftus, E.F., Altman, D. and Geballe, R. (1975) Effects of Questioning Upon a Witness' Later Recollections, *Journal of Police Science and Administration* 3 (2): 162–165.

Loftus, E.F., Donders, K., Hoffman, H.G. and Schooler, J.W. (1989) Creating New Memories That Are Quickly Accessed and Confidently Held, *Memory and Cognition* 17 (5): 607–616.

Loftus, E.F. and Hoffman, H.G. (1989) Misinformation and Memory: The Creation of New Memories, *Journal of Experimental Psychology: General* 118 (1): 100–104.

Lynch, M. (1998) The Discursive Production of Uncertainty: The OJ Simpson "Dream Team" and the Sociology of Knowledge Machine, *Social Studies of Science* 28 (5/6): 829–868.

Rattner, A. (1988) Convicted but Innocent: Wrongful Conviction and the Criminal Justice System, *Law and Human Behavior* 12 (3): 283–293.

Sear, L. and Williamson, T. (1999) British and American Interrogation Strategies, pp. 65–81 in D. Canter and L. Alison (eds.) *Interviewing and Deception*, Aldershot: Ashgate.

Skolnick, J.H. and Leo, R.A. (1992) The Ethics of Deceptive Interrogation, *Criminal Justice Ethics* 11 (1): 3–12.

Thomas III, G.C. and Leo, R.A. (2012) *Confessions of Guilt: From Torture to Miranda and Beyond*, Oxford: Oxford University Press.

Weber, S. (2008) The Hidden Truth: A Sociological History of Lie Detection (PhD Thesis), London: London School of Economics.

Chapter 7

Subjects of suspicion

> Sex offending is shrouded in secrecy and denial. . . . Use of the polygraph with sex offenders under community supervision can serve to help the probationer overcome denial, aid in developing treatment plans, be a deterrent for sexual reoffending and assist in monitoring compliance with the conditions of supervision.[1]

For as long as there have been lie detection practices there have been subjects of lie detection: those social actors most intensely targeted by the powerful organisations which use torture and technology to extract confessions. As the quotation above shows, sex offenders are one recent example of such a group but there is a long history of subjects of suspicion. These kinds of subjects become targets according to cultural norms which construct them as 'Others' and as being dangerously so on a scale requiring significant interventions. Considering the truism that sex and lies go hand-in-hand, it is little surprise that women and homosexuals have, unfairly, become key subjects of lie detection practices at one time or another. Each social era has had its subjects of suspicion, who – through the escalatory politics of moral panics – can become cast as monstrous liars. Catholics and communists have been subjected to lie detection, but are no longer key political targets. Today, it is terrorists, especially Muslim terrorists (and through racial generalisation Muslims more generally) who have come into focus. A further party, the paedophile or sex offender, has also become a central target of scrutiny with the polygraph machine. Although modern readers might rightly discern a difference between women and sex offenders being made subject to lie detection, the former being innocent of monstrosity and the latter guilty, there has always been a degree of Othering in the construction of lie detection's primary targets, and some women have indeed been deemed monstrous at points in Western history. In this regard, the subjects of lie detection upon which I place an emphasis – women, homosexuals and paedophiles – have in common their construction as specifically *sexual* monsters, with important implications for how it is that torture and technology come to be seen as possible antidotes to the social ills which such monsters come to embody.

Although deception and fabrication are generally treated as a problem for social organisation, I will argue that the process of Othering marks certain groups as different in ways which make their lies especially troubling to socio-political order, thereby justifying a range of often more brutal practices for managing mendacity. The idea that certain kinds of people are so fundamentally different or deviant as to make them monstrous in some way forms part of several sociological approaches, including work on 'folk devils' and 'moral panics' (Cohen, 2002) and governmental analyses of abnormality and deviance (Foucault, 2003; Rose, 2000). Each is also concerned with the ways in which risk is managed in society. Drawing on literature from these two broad approaches will prove instructive for understanding the use of lie detection practices (whether in the form of torture or technology) in the interrogation and control of sexual monsters.

Rose (2000: 333) summarises the way in which risk is managed in the contemporary era:

> [T]he identification, assessment, elimination or reduction of the possibility of incurring misfortune or loss has thus become an integral part of the professional responsibility of a host of professionals. The respecification of one dimension of the problem of control in terms of risk is bound to a revised governmental role for such professionals, to manage dangerous sites and dangerous persons on the territory of the community.

Amongst such professionals, the polygraph examiner has been enrolled, through governance of medical and judicial conduct, into practices for controlling monstrous groups over the course of the twentieth century. Sex has been crucial to this arrangement.

Foucault's (2003) analysis of abnormality is instructive here, for it is closely concerned with sexuality and gender. He argues that the modern psychiatric and judicial understanding of sexual abnormality was constituted through the development and then fusion of three abnormal figures in the centuries leading up to the 1900s: the human monster, the individual to be corrected (ITBC), and the onanist, or masturbator. The first of these, the human monster, violated all laws, including the laws of nature, and was best characterised by the figure of the paedophile. As psychiatry and other human sciences developed, the three figures fused, bringing together a range of techniques of power to constitute a more coherent technology of abnormality at the intersection of political governance, juridical practices, medical knowledge, family relations and disciplinary institutions. It is against this background that the paedophile became the focus of a pervasive moral panic in the West.

Whilst the notion of moral panic is less firmly connected to the constructs of sexuality, Cossins provides a broad argument which links the periodic emergence of folk devils to gendered bodies:

> the construction of a folk devil – the indispensable step in the development of a moral panic – was a product of cultural and historical sexing processes of

> either the male or the female body, with its characteristics derived from the essentialist, negative values associated with maleness or femaleness.
>
> (Cossins, 2015: 258)

There are numerous examples of sex, gender and sexuality playing a role in the emergence and intensification of moral panics (Fahs et al., 2013; Fejes, 2016). In a similar fashion, I draw attention to the role which the human body plays in bridging gender, sex, monstrosity, morality, risk and public panic to reflect upon how lie detection has been embroiled in a messy and often violent political history of producing and managing difference and deviance. I also provide the first detailed account of the movement of the polygraph machine into the United Kingdom, where lie detection has historically found little room to take root. I argue that we can understand the use of the polygraph in the context of sex offender management through the established literatures, for there is a degree of contiguity with the past. However, I also argue that the management of sex offenders presents a key discontinuity in the polygraph's tale, as it is used not only as a form of interrogation but also of therapy. Whilst the polygraph has always traded on its connections to physiology and science; in the context of paedophilia, its mode of operation becomes more explicitly framed by clinical discourse, and the examiner's role not only resembles that of a doctor but comes dangerously close to having the examiner take on such responsibilities, for the polygraph is a fundamental part of the contemporary treatment framework.

The subjects of lie detection

Women have long been constructed as the deceptive sex. This is due in part to the biblical representation of women which has powerfully framed the West's interpretation of femininity. Lying is a social characteristic often used to depict women in important Christian parables and teachings. As Fuchs (1985: 137) illustrates:

> It is a motif that runs through most narratives involving women, both condemnatory and laudatory ones. From Eve to Esther, from Rebekah to Ruth, the characterization of women presents deceptiveness as an almost inescapable feature of femininity.

The construction of women as the deceptive sex is reflected in the undue attention paid to them in the development and use of lie detection methods. For example, in the periods in which torture functioned as the most common means of lie detection, the politics of truth and lying were explicitly gendered. This was also mirrored in emerging scientific practices which were deeply entangled with torture (see Chapter 2). Katharine Park's (2006) work on the links between secrecy and women's bodies in the Enlightenment is exemplary in this regard, for it shows that women were bearers of secrets about life and reproduction, passed down through generations, but which were collected and published by male scientists who sought, through dissection of the female anatomy, to dispel the mysteries of nature. Representations of female anatomy derived from dissection made use of the emerging modes of signifying objectivity in print (Daston and Galison, 1992) and sometimes also drew upon the erotic mode of artistic representation which positioned women as passive objects to the male, scientific gaze (Merchant, 2008).

Women have also been the principal victims in the horror of recurrent witch trials around the world. In Europe and America both men and women were tried as witches (Geis and Bunn, 1997), though women were the primary focus and it is the witch as woman that survives in the cultural imagination of the West (Rowe and Cavender, 1991).

The idea of the 'diabolical witch', a person using magic and satanic powers to inflict harm on others, developed in the 1400s and spread through Europe over the following three centuries. This coincided with great shifts in urban living: the proportion of strangers, outsiders or immigrants living within towns and cities had begun to increase. The forms of criminal prosecution tended to be structured according to one's reputation and level of integration, so that the poor and outsiders were more likely to suffer at the hands of torturers and received the greatest punishments. The diabolic witch represented the worst kind of outsider because she was much harder to identify, often lacking in ostensible foreignness or difference. Satanic beliefs and secret magic made her an evil Other who lived on the inside of urban communities, rather than on its visible fringes. The prosecution of witches therefore often involved much more severe forms of torture (Stokes, 2011).

Since women were the most common targets in the emergence and spread of witch trials, the use of torture not only for their punishment but also for investigative purposes reflects the early enrolment of lie detection methods into broader inequalities suffered by women throughout history, and evidences the integration of lie detection practices with power relations which typify patriarchy.

By the late 1800s, fear of witches in Europe and America had declined but the construction of women as irrational, deceptive and untrustworthy persisted. For most of the history of the social and natural sciences, women have been conceptualised as natural-born liars and manipulators (Denery II, 2015). Such accounts often found women's deceptiveness to be a fault of their biological nature, and so it was theirs rather than men's bodies which became the main subject of much psychological and criminal investigation during the Enlightenment and in the growth of criminology at the turn of the twentieth century.

> Late nineteenth-century criminology codified the female body as a cipher, a bearer of secrets about criminality. . . . The science came to regard the female body as a conundrum, nothing less than the numinous key to the puzzle of criminality.
>
> (Bunn, 2012: 52)

Embodying the irrational, selfish and criminal, the woman figured as the ultimate liar for the scientists of the period, which deeply informed the political framing of the lie detector and its media persona:

> [I]t is hardly surprising that the pioneers of the polygraph (and newspaper editors quick to sense the "hot" angle to their story) used gender typing to convey the "objective" nature of lie detection techniques and dramatize the ability of polygraph operators to ferret out hidden thoughts.
>
> (Alder, 1998: 498)

Women often featured in the advertisements which promoted the polygraph machine, and they played a key role not only as rewards for men, but also as adversaries (in the form of the femme fatale, for example) for the stereotypically masculine heroes of detective fiction, all of which was instrumental in promulgating the myth of lie detection (Littlefield, 2011). Even more telling is how women's bodies figured in prostitution, which was – in the late 1800s – the ultimate 'social evil, surpassing drunkenness, blasphemy, and adultery' (Bunn, 2012: 62). That women might use their bodies for profit was – it seems – the ultimate deception and economic manipulation of heterosexual men's desires.

The intensification of scientific studies of women's bodies meant a growing interest in the physiology of the emotions, which were understood to produce measurable effects in the body and thus be amenable to positivist study, opening up the door to lie detection through mechanical instrumentation. As Bunn (2012: 74) summarises: 'One of the conditions for the lie detector's emergence was, therefore, the conceptualisation of criminal woman as the quintessential problem of criminology' – a problem which could be understood better through the application of mechanical devices for the study of their emotions and psychology. The lie detector's development in the 1920s and 30s promised to help men finally win the battle of truth in the war of the sexes.

Though the *Frye* decision was an early test of the lie detector's potential in the scientific fight for truth, and appeared to be a failure in the criminological adventure, conversely it foreshadowed a massive expansion in use of the device, especially during McCarthyism and into the long decades of the Cold War. In this climate of suspicion, coupled with concerns of serious intelligence infiltration by foreign agents, the polygraph was propelled front and centre as a tool with which to defend not only the security of the United States and the integrity of the intelligence services but also the core of American identity.

Communists were now constructed as a potent threat to the purity of the USA. In tandem, women were portrayed 'as passive victims, anguished wives, or caregivers' in the war. Men were either war heroes, diligent patriots or spies both foreign and domestic. As Sharp (2000: 129) reminds us: 'This should not come as a surprise, for in traditional national rhetoric, men are metonyms of the nation in that the nation is embodied in each man and each man comes to embody the nation.' As the United States' individualist, capitalist imagined community came under threat, an image of the manipulative and secretive communist was created through political speeches, corporate news media and even in everyday magazines such as *Reader's Digest* (Sharp, 2000), making the communist spy a monster and the new master of deception. As such, women were no longer the key target of lie detection practices, for communist spies had become the key liars to uncover.

This shift in focus resulted in the broad introduction of the polygraph throughout US institutions, but most recognisably in those security contexts which seemed vulnerable to espionage. In 1947 the Atomic Energy Commission became the first government agency to engage the machine (Cook and Robinson, 1947). In the decades which followed, the Departments of Energy and Defense, the National Security Agency and a range of other large branches of the US government bearing responsibilities for (inter)national security adopted the device for employee screening processes and internal investigations.

For all the talk of communist barbarism, this period also saw the official return of torture in US security. The psychological phase of the third degree and experiments with truth technologies like the polygraph machine had helped to remove (for the most part) physical torture from police interrogation (Leo, 2008). However, this spurred a series of experiments with bodily and psychological torments that could not easily be evidenced and could be sustained for much longer periods. Although the police played a role here, the Central Intelligence Agency (CIA) was chiefly responsible for the innovations in torture which we live with today: its scientists and interrogators developed a range of new and perhaps even more cruel means of extracting information from suspected communist spies and sympathisers, just as the use of the polygraph skyrocketed.

> From 1951 to 1962, the CIA led a secret effort to unlock the mysteries of the human mind, searching for methods that would make spies reveal their secrets, turn enemy agents into double agents, and persuade millions through a subtle psychological warfare. With costs that peaked at a billion dollars a year, this psywar project developed, as its most lasting legacy, a distinctive method of psychological torture.
>
> (McCoy, 2007: 16)

Here, the lie detector played a dual role: it was both practical aid to torture of suspects, and political cover for the security services. As a public face of the Cold War investigations, the polygraph machine's power to root out spies emphasised American technological superiority and civility, but it also masked the CIA's more covert developments in psychological torture.

Despite its increased deployment, the lie detector proved to be of little use in fighting the communist threat. Very few spies were discovered through its application. Fortunately for polygraph advocates there was also a sexual monster to be confronted. In the forties and early fifties, the US State Department had come to be seen as a haven for sexual deviants. In what would become known as the 'Lavender Scare' (Johnson, 2004: 18), the counterpart of McCarthyism's

'Red Scare,' the chairman of the Republican National Committee claimed that 'sexual perverts' were '[p]erhaps as dangerous as the actual Communists'. The polygraph was thus instead adopted to play a major role in the interrogative procedures of the ominously titled 'Miscellaneous M Group', an investigative unit formed to handle any type of moral deviation in the State Department, though the focus was overwhelmingly on homosexuality.

> All male applicants were subject to a personal interview by security personnel who specialized in uncovering homosexuals. If suspicions were raised, the applicant would be given a lie detector test. Security officials, for example, suspected an applicant in 1947 of homosexuality because of his "mannerisms and appearance (use of perfume, etc.)." He was subsequently given a polygraph exam, confessed to homosexual activity, and was rejected
>
> (Johnson, 2004: 73).

It was not only applicants for positions who were being screened but also existing employees. In 1953 the Group achieved a 'success rate' of forcing one resignation of a homosexual employee 'every two-to-three days' (Johnson, 2004: 128). In a memoir on his work as a CIA polygraph examiner during this period, Sullivan (2007: 22) recalls that 'CIA examiners over the years probably obtained more disqualifying information regarding homosexual activity than any other issue. . . . With the exception of criminal activity . . . most employees who lost their clearances lost them as a result of admissions of homosexual activity.' Another former examiner laments: 'The hunt for homosexuals was indeed vigorous. I had a number of male and female friends who suddenly disappeared from the Agency. I would see them one day and the next day they were gone' (Chapman, 2008: 4).

Rather than being a scandal in the press, the media's coverage of homosexuality in the US government followed the pattern of a moral panic more than it did a reflection of the dire inequalities which gays faced in the workplace. When stories were fed to the press about men being removed from their jobs, it only stoked the fires and propelled further calls for action to be taken on the problem of homosexuality. The Attorney General, the Secretary of State and the Director of the FBI (J. Edgar Hoover) all supported the use of the polygraph in extracting gays from government (Johnson, 2004: 145). The line between the fear of communists and the fear of homosexuality was difficult to draw. In America, therefore, the general anxiety induced by the Cold War was reflected in an intensification of suspicion of not only political but also sexual Others.

Similarities between the experiences of women and gay people can thus be drawn. As with women, homosexuals have suffered a history of Othering in which they have been characterised as abominations through Christian teaching but also in secular philosophy and politics (Bayer, 1981). Much as women became subject to scientific scrutiny, so too have lesbian, gay and bisexual people. By the time of the Lavender Scare, gay people had been classified as mentally ill by American psychiatry, meaning that they were not only morally deviant and unnatural but also physically and mentally abnormal in ways which warranted study and treatment. Women too were seen to be morally and mentally inferior to rational men and in need of analysis.

It was in this period that homosexuality also grew to be conceptualised less as an action and more as an identity position or way of being. In other words, as gays became political subjects in America they also became subject to lie detection through a moral panic regarding their mendacity. Lying and its management are thus thoroughly entangled with the rise of panics about such social, sexual monsters. Through the socio-political construction of certain morally and physically monstrous groups as deceptive and untrustworthy, lie detection practices are positioned as technical solutions by virtue of the way in which they work. It is because of their claims to seeing the truth in the human body, in the form of either torture or polygraphy, that they take on this kind of role. The body becomes the switch-point in a power relation which makes truth inaccessible through other means. The construction of women as witches and gays as abominations means that the usual relations of trust (in the form of truthful speech, sameness, etc.) are not used. The lie detector frequently finds itself at home in these scenarios: as vulnerable groups face the most desperate of situations, it grows in power.

We can read a further and more general pattern from these examples of how women, communists and homosexuals have become subjects of lie detection, one which recurs throughout the history of the use of torture and lie detection practices. The most feared targets are always those outsiders whose difference is most difficult to discern by other means. What makes them so monstrous is not only their stigmatised characteristics or actions, not merely that they cannot be trusted, but also that they bear too much similarity to 'normal' people.

Of significance here is Foucault's (2003) argument regarding the judicial and medical figure of the 'individual to be corrected' (ITBC), which exemplified the governance of minor, everyday aberrations of social life. Each person had to be guided back to normality by social institutions and the professionals who inhabited them. But as the figure of the ITBC began to fuse with the human monster in psychiatry and the psychological sciences, the petty deviations of the ITBC came to be treated as warning signs of a deeper problem, which, if not corrected, risked the individual transforming into something much more heinous. The same happened with masturbation, which became the root 'of almost every possible evil' (Foucault, 2003: 59), the 'explanatory principle of pathological singularity' (Foucault, 2003: 60). The positive outcome of this for psychiatrists was that monsters could now be studied. They were normal people gone bad through excessive masturbation and failed correction of deviance. The problem was that now each person also posed a potential risk if he or she was allowed to slip beyond minor aberrations.

From the twentieth century, sexual monsters thereby came to represent a looming threat inside normality, inside each individual sexual actor, but also – and more imperceptibly – troubled the social codes of truth production. If monsters can be hidden amongst us, the mechanisms which we use to trust people are thrown into doubt. As Bauman (2004: 58) has argued in a different context, this fear often breeds violent reactions:

> [They] are not feared and hated for being different – but for being *not different enough*, mixing too easily into the crowd. Violence is needed to make them spectacularly, unmistakably, blatantly different. Then by destroying them one could hopefully eliminate the polluting agent that blurred the distinctions and thereby recreate an orderly world in which everyone knows who they are and identities are no longer frail, uncertain and precarious.

Through the figure of the monster, lie detection has played a role in a complex and shifting arrangement of torture, criminal investigation, prosecution and punishment, and has often come into play through its promise to solve socio-political crises of confidence in social trust and truth practices, channelled through the monstrous targets which such panics help to create.

The spread of polygraph sex offender testing in America

In the 1960s polygraph admissibility decisions had already begun to diversify – new arguments were made for the admission or exclusion of lie detector evidence. By this time, the machine had spread across the US and taken root in a range of situations. It was in this period that the courts, in collaboration with probation officers, medical professionals and polygraph experts, instituted a new way of applying the lie detector to problems of truth: to extract information from known sex offenders to better manage the risks they posed. This use of the polygraph in sex offender treatment programmes remains one of the most understudied turns in the story of lie detection technology, but it is one of the most important shifts in the polygraph machine's history, for it has greatly expanded the use of the device within probation contexts, helped chart a course to implementation in the United Kingdom and opened up a new conceptual framework for its application.

The seeds of the approach germinated in 1966 when Judge Clarence Partee of Illinois began to use the polygraph to elicit confessions of other crimes as he was sentencing criminals in his court. Around the same time, Judge John Tuttle of Washington began to require some probationers to take periodic polygraph examinations (Shaw, 2002–2003) to check up on their adherence to probation conditions. Neither of these shoots took root in the justice system, but they prepared the ground for sex offender testing.

There are three uses of the polygraph machine in sex offender probation programmes. First, it is used as an investigative tool in much the same manner as it is in police stations across the USA. During probation, concerns about possible breaches of conditions might arise, for example because of disclosures during therapy or because of reports from the community, probation officers or other parolees. Worries might also arise that the offender is guilty of a crime that has been recently reported which matches his offending pattern. When such concerns are raised, a 'single incident examination' is conducted with offenders, for example to assess whether they are guilty of the new assault or a breach of their probation conditions. Second, the machine is used as part of the surveillance apparatus. Offenders must submit to (and often pay for) regular examinations (for instance taking one test every six months) to check that they are fulfilling their probationary criteria, e.g. they are not interacting with past or potential victims, they are not maintaining contact with other offenders, they have not visited banned locations. This is a relatively novel implementation of the device, though there are existing contexts in which the polygraph has been used as a regular surveillance test (e.g. in employee screenings in business and government departments, or in the regular testing of police officers). Third, the polygraph forms part of the therapeutic process and is used to help to produce information about the offenders' desires, beliefs, intentions, past actions, dreams and so on. It is this latter application which represents the most novel development.

In the late 1980s Oregon became the first state systematically to implement the polygraph as a tool in the management and treatment of sex offenders on probation.[2] But it was not until the 1990s that the US witnessed the gradual spread of lie detection examinations into probationary programmes. In 1995, the Tennessee legislature passed a law that made polygraph testing mandatory for sex offender probationers[3]; Colorado followed suit in 1996.[4] Hawai'i was also using the technique before the turn of the century, as were the states of Florida, Illinois and Kansas.[5] Along with these early adopters, several other states, including some of the most populous, have been instrumental in the development of guidelines for the use of the machine with sex offenders; for example, California,[6] Texas,[7] Indiana,[8] New Mexico[9] and South Carolina[10] have all helped to develop a more coherent framework. An idea which had developed in the 1980s had only limited success for its first decade but by the mid-90s, as states began to produce official guidelines and implement new legislation, it had already spread to around 30 per cent of individual probation programmes, escalating to 60 per cent by 2000 and then rising further to 70 per cent in 2002 and 79 per cent in 2009 (McGrath et al., 2010).

To begin with, the use of the machine in this context was rather haphazard, with very little underpinning research or philosophy. Towards the end of the 1980s, however, a new strategy, called the 'containment approach', was developed to monitor, manage and treat sexual offenders, which more formally conceived a role for the lie detector. So although the polygraph's use appears to have preceded the organised development of the containment approach, it was this framework which helped to legitimise and spread the lie detector through post-conviction probation programmes in the 1990s.

The containment approach synthesised, standardised and provided a conceptual structure to underpin a variety of pre-existing monitoring, management and treatment strategies, including polygraphy. The emphasis on 'containment' was a direct response to the release on probation of a greater number of sex offenders into communities, which itself was a reflection of the growing crisis in US prison populations. In 1980 there were an estimated 20,500 sex offenders in state prisons, but this had quadrupled to around 88,000 by the mid-90s (English et al., 1997). This was a drop in the ocean, however, compared with the broader problem of prison overpopulation.

Release of offenders into the community became more common as the prison crisis grew. In the mid-90s there were around 230,000 offenders convicted of rape or other sexual assaults living under the care, custody or control of corrections agencies, with approximately 60 per cent of these offenders being 'contained' within the community (Greenfeld, 1997). It has long been established that sex offenders have high rates of recidivism (for a meta-review, see Furby et al., 1989). At the same time, victims of sexual offences have demonstrated very low rates of reporting, meaning it is difficult to estimate quite how significant sexual crimes are within the context of crime as a whole. In the mid-90s, for example, only 32 per cent of offences were being reported (Greenfeld, 1997), a problem which continues to this day as the same proportion of rapes and other sexual assaults were reported in 2015.[11]

These statistics were troubling not only because they evidenced an uncertainty in the impact of sex offenders living in the community, but also because it indicated a problem in how victims of sexual crimes were being treated by the police. Probation and parole officers working with sex offenders in the late 1980s and early 90s had become frustrated with the existing guidelines and therapeutic interventions, which tended not only to treat sexual crimes in the same fashion as any other crime but also to implicitly blame victims as much as they did offenders. As English et al. (2016: 714) reflect:

> The containment approach grew out of the frustration of probation and parole officers who were constrained by ineffective policies and procedures resulting from social attitudes that minimized the crime. In the same way that intoxication was once viewed by the court as a mitigating rather than aggravating circumstance, sexual abuse was seen as a family problem or a simple misstep by the perpetrator.

In this regard, the construction and treatment of sexual offenders before the development of the containment approach tended to emphasise individual choices gone awry. Sex offenders were everyday men who had succumbed to unchallenged desires.

However, the growth of the containment approach coincided with and became politically salient by virtue of several media-driven scandals regarding crimes against children, which spurred a significant moral panic in the early 1980s about the abduction and murder of children by strangers. Over the next two decades, paedophilia became a prominent theme in the US news media and came to stand in for sexual crimes as a whole. The media sensationalised tragic cases of child abduction, molestation and murder

> during the late 1980s and early 1990s with the 1987 case of Earl Shriner raping and mutilating a six-year-old boy, the 1990 disappearance of Jacob Wetterling in Minnesota, the 1993 abduction and murder of Polly Klaas in California, and the sexual molestation and murder of Megan Kanka in 1994 in New Jersey.
>
> (Neuilly and Zgoba, 2006)

Campaigns, often mounted by grieving parents, resulted in the introduction of new federal laws governing the investigation of missing children (Neuilly and Zgoba, 2006; Critcher, 2008). Although the numbers of offenders in prison and in the community seemed to be increasing, sexual offences still represented a very small proportion of overall crime, and there is some evidence that the percentage of sex offences was actually decreasing as the moral panic grew (Neuilly and Zgoba, 2006). As a result, sex offenders became modern folk devils in the 1990s, inhabiting the centre of a broader moral panic about the care of children and the meaning of childhood (West, 2000; Critcher, 2002; Pratt, 2002; Neuilly and Zgoba, 2006).

The mid-90s was indeed decisive: the US Congress enacted the 1994 Violent Crime Control and Law Enforcement Act,[12] which included the 'Jacob Wetterling Crimes Against Children and Sexually Violent Offender Registration Act'.[13] This was amended two years later in the more widely known 'Megan's Law'. Together, these acts implemented the federal (and subsequent state) policies requiring sex offenders to be registered in public databases and for communities to be notified when offenders moved into an area. These changes singled out sexual offenders as different kinds of criminals, posing a different kind of risk and therefore requiring different kinds of management.

Set against the background of scandals, public outcry and the growing number of sex offenders living within American communities, alongside the challenge of recidivism rates, the troubling underreporting of crime, and newly developed legislation, the Assistant Attorney General convened a 'National Summit Promoting Public Safety Through the Effective Management of Sex Offenders in the Community' in 1996. This large meeting of key actors from across the USA helped to galvanise the containment approach as the default mode for managing the offender population. As such, the polygraph had become embroiled within a broader socio-political shift in American governance of children and sexual crimes. It provided examiners with new entrepreneurial opportunities but also with a chance to reconstruct lie detection practices for use in a new context.

The spread of the polygraph to the United Kingdom

It was in similar socio-political waters, but a decade later, that the polygraph put down its first anchor in the United Kingdom. Historically, lie detection has failed to find a government organisation or a community of professionals in the UK willing to adopt it, despite occasional government explorations of its potential.

Although it was briefly examined in the 1960s,[14] the machine was given its first serious consideration in the 1980s. Originally, it was examined as part of a report on the review of criminal procedure by the Royal Commission (which had begun in 1978).[15] Commission officials travelled to St. Louis (Missouri) and Columbus and Cincinnati (Ohio) to meet with solicitors, chairpersons of police authorities and police officers to discuss their use of the polygraph machine in police interrogations. The report concluded that American officers found the lie detector to be a useful aid in their investigations but that a distinct lack of certainty regarding its scientific status should preclude experiments with the device in Britain.

Second, the polygraph was considered as a possible deterrent to Soviet espionage in the last years of the Cold War, paralleling some of the security concerns which still dominated the US government's agenda. During an evaluation of intelligence practices following the case of Geoffrey Prime, a former British spy who divulged secrets to the Soviet Union in the 1970s, members of the British judiciary and security services visited the United States to meet with security officials from all three of the US government's intelligence agencies.[16] As part of their conversations, they found that Prime had apparently avoided routes into security positions which might have required him to undergo a polygraph exam; for example, on US business. Taking this to be evidence that the lie detector was a potential means for preventing such espionage in the future, the Commission's report recommended a pilot of the device in British intelligence services, most importantly within the Government Communications Head Quarters (GCHQ) where Prime had worked.[17]

It was a swift change in direction compared with the Royal Commission's position, which seems in part to have been a response to international pressure. Sir Robert Armstrong, the Cabinet Secretary, told civil service union leaders representing concerned staff at GCHQ that the US National Security Agency had questioned whether to continue sharing intelligence with Britain if the government refused outright to test the feasibility of the polygraph. The union was not convinced, but the House of Commons Employment Committee stood fast and arranged an independent assessment of a pilot scheme to be implemented at GCHQ in summer 1984. The union responded by threatening disruptive action and to take the government to court if any GCHQ staff member was affected by the results of a lie detector test (Arthurs, 1985).

Although it related to one of the most sensitive areas of British politics, the battle over GCHQ was amongst the smaller ones fought by Margaret Thatcher's government in its war against the unions. The Conservatives decided that from 1 March 1984 GCHQ staff would no longer be permitted to unionise. It was a vicious response to GCHQ's worries over staff surveillance, but one which paled in comparison to the storm breaking in the miners' strike that month. The issue was fought in the House of Commons,[18] where Conservative minister Sir Geoffrey Howe (Secretary of State for Foreign and Commonwealth Affairs) tried to play it off as a battle over pay rather than the polygraph, painting unionisation not only as a political and social threat but as a matter of life and death when it came to national security. Labour MP and former Secretary of State for Defence Mr Denis Healey was not taken with the characterisation and claimed the government had been conned into adopting an 'American technological gimmick'.[19] In the end, the case was brought before the House of Lords, which found that the government was within its rights to ban unionisation of GCHQ staff and to introduce the polygraph machine without notice.[20]

As the miners' strike intensified, the UK media had bigger political stories to cover, but the Americans forcing the polygraph on British spooks did receive some exposure in *The Guardian*,[21] and popular science magazines ridiculed the Conservatives' insistence on introducing the device, citing the American government's own reports showing that the polygraph did not work.[22] Official scientific journals also played a role in contesting the claim that the polygraph could assist in counter-intelligence. The British Psychological Society produced a highly critical review, concluding:

> There is fundamental concern about the extremely high proportion of innocent (honest) people who will be classified as deceptive by the polygraph. Some studies have shown that the detection rate is no better than chance – in other words, the toss of a coin would have been as effective in identifying innocent people.[23]

Eventually, the combination of scientific and political resistance curtailed plans to introduce the polygraph machine, though GCHQ staff still lost their right to unionise, meaning that the threat of the lie detector played an unusual role in the restructuring of the espionage community under Thatcher, even if the machine was not once put into use for employee screening.

It was not until the Conservatives were toppled by Tony Blair in the New Labour years that the polygraph received a second try at British integration. A variety of measures were introduced in the early years of the Labour administration that took office in 1997 and are now to be found in the Criminal Justice Act 2003 (CJA 2003) and Sexual Offences Act 2003 (SOA 2003). The CJA 2003 provides for extended periods, of up to eight years, of supervision following release on licence from prison for persons convicted of specified sexual offences.[24] These laws were implemented as part of a game of one-upmanship between the Tories and Labour in the media, with each party trying to outdo the other in how tough it could appear to be on crime (Young, 2003; Newburn, 2007). Sex offenders became key political pieces in this game, since they could be used to propose practically any punishment without segments of the press or public crying foul. This was, in part, because of their long-term stigmatisation, but also because the rhetoric of 'tough on crime' punishment was combined with the rise of British moral panic regarding sex offenders that mirrored the American situation. As Critcher (2002: 521) summarises:

> In the summer of 2000, a mass circulation British newspaper, the News of the World, conducted a campaign to "name and shame" convicted paedophiles. Initially attacked by other newspapers and authoritative agencies, the campaign gained momentum, despite provoking vigilante activity. Negotiations between the paper and influential pressure groups produced a "charter" of action presented to the government. Virtually all its procedural reforms were conceded.

So it was that the advocates of the polygraph machine got their best shot at success. By now the containment approach had spread across the USA and had helped to integrate polygraphy into post-conviction sex offender programmes. There was a developing scientific literature and the American Polygraph Association had started to professionalise the 'examiner as therapist' role. A few respected psychiatrists and scientists in the UK took it upon themselves to begin their own research into lie detection, supporting and helping to implement a government pilot of polygraphy with sex offenders under their care. In a volatile political and media climate, the Labour Party had promised to examine the potential of a British counterpart to America's programme, and now began to legislate for it.

Sections 28 and 29 of the Offender Management Act 2007 were introduced to allow for the attachment of a 'polygraph condition' to the terms of the release from prison on licence of sex offenders.[25] This meant a scientific trial could be conducted and a pilot of the lie detector in paedophile probation programmes was quickly announced. Sections 28 and 29 were brought into force early in 2009 and the pilot began in April 2009, ending in 2012.[26] It was deemed a success by the research team and by the Labour administration, though its broader implementation was handed over to the coalition government of the Conservative Party and Liberal Democrats, who were only too happy to take credit for its success and its demonstration of their intolerance of crime. As Justice Minister Jeremy Wright[27] pronounced when rolling out the polygraph across the UK, 'Introducing lie detector tests, alongside the sex offenders register and close monitoring in the community, will give us one of the toughest approaches in the world to managing this group.'

At almost the same time the British Psychological Society released another report on lie detection, which was equally disparaging as their first had been two decades prior:

> The vast majority of published studies on polygraphic lie detection have been laboratory-based and they have not been that successful in recreating many of the factors that would be present in real life (in the 'field') such as high costs of being judged as lying. Thus serious questions persist about generalising beyond laboratory situations.[28]

But on this occasion, it was to no avail. The polygraph programme was legislated and paid for, and it was quickly introduced to probation teams across the country. This was the first time the lie detector had been taken seriously by a British institution. The paedophile had become the latest monster to help frame lie detection as a technological fix for a socio-political problem in the United States and then in the United Kingdom. Although it was not quite so simply the case that the Americans had 'forced' the lie detector on the British government in the case of GCHQ, when it came to sex offenders it was manifestly not the Americans who were to blame. American governance had helped make it possible, certainly, but it was turn-of-the-century British politics which – in the end – provided the monster and the means required to justify the adoption of a technology which had been so successfully resisted for the previous four decades.

The polygraph in the containment approach

At its heart, the containment approach is a way of understanding sexual offenders, and is targeted most powerfully at paedophiles rather than at other sex offenders. It focuses on the risks they pose and it is a victim-centred approach but one which also seeks to understand the sexual offender as an individual rather than as a generic problem to be dealt with in a uniform manner. It is an integral part of the containment approach that each offender be given a personalised therapeutic programme, tailored to his specific crimes, desires and denials. At the same time, however, the framework posits key personality features which sex offenders tend to exhibit and depends upon this typical characterisation for its legitimacy and efficacy. The approach acknowledges that the majority of sexual offenders are men: in the USA 99 in 100 are male (Greenfeld, 1997). It also understands such men to be secretive, deceptive and manipulative. As early advocates of the approach argued in one of their first reports on the matter:

> Sex crimes flourish in secrecy. Sex offenders have secretive and manipulative lifestyles, and many of their sexual assaults are so well planned that they appear to occur without forethought. The skills used to manipulate victims have also been employed to manipulate criminal justice officials. Many sex offenders are otherwise highly functioning people who use their social skills to commit their crimes. Sex offenders typically have developed complicated and persistent psychological and social systems constructed to assist them in denying and minimizing the harm they inflict on others, and often they are very accomplished at presenting to others a facade designed to hide the truth about themselves.
>
> (English et al., 1997: 2)

This account is a typical one, and evidences the various ways in which sexual offenders are characterised in relation to lying, secrecy and deception. First, they are seen to live a secret life, one which is kept from intimate others and is a learned response both to the criminal implications of their desires but also to their stigmatisation in society. Second, the creation of a secret life needs constant management, and sex offenders are consequently assumed to be masters of secrecy and lying, skills which enable them to manipulate their families and friends, victims, criminal investigators, social workers and treatment professionals. Third, they are skilled at impression management, making it difficult to discern their deviance and criminal acts and allowing them to blend in with others. As the probation conditions of Deschutes County in Oregon put it, 'Sex offenders are known to be manipulative and secretive; their outward appearance and conduct does not usually reveal their sexual deviancy and/or criminality.'[29] Finally, sex offenders are understood to be self-deceptive, or – to put it in the more common therapeutic frame – to be deeply in denial about the factual nature of their desires and crimes, or about the moral status of their actions and intentions.

Characterising sex offenders as masters of manipulation, even of themselves, means that their statements about their thoughts and feelings cannot be trusted, but it is through talking to offenders that probation officers and therapists develop a personalised treatment programme and seek to reduce the risks that such men pose. Beech et al. (2003: 340) outline this difficulty in the UK context:

> Any assessment of a sex offender should include . . . information about the antecedents, the behaviors, and the consequences of offending (the ABC model). This should include the actual behaviors carried out, along with the accompanying thoughts and emotions. Unfortunately, this is not always a straightforward task with sex offenders because they are frequently in some degree of denial about aspects of the offense and are therefore not willing to be completely truthful about the information that the assessor needs to obtain.

The polygraph machine thus plays a powerful role, and whilst early papers cautioned against use of the device in this context (Becker and Quinsey, 1993) it rapidly became an integral part of the US and UK approaches by acting as a tool for extracting confessions which could be given greater credence. As (English et al., 1997: 4) explain, polygraph examinations are used:

> to obtain complete sexual history information and to monitor the offender's deviant fantasies and external behaviors – particularly access to victims. Data obtained during the polygraph examination provide vital management and compliance feedback to the treatment provider and probation/parole officer.

Obtaining a sexual history is a standard procedure in the treatment of sex offenders. According to the Arizona state guidelines, the kinds of things that are investigated during a sexual history exam include:

> Age of onset of expected normal behaviors . . . degree of use of pornography, phone Sex, cable, video, or internet for sexual purposes, current and past

> range of sexual behavior . . . thoughts [and fantasies] preceding and following crimes; . . . masturbation, use of tools, utensils, food, clothing, current sexual practices . . . motivation to change . . . attitudes [including sexual] toward women, men, children . . . level of denial, level of deception.[30]

An interview regarding these behaviours, thoughts, fantasies, motivations, attitudes and deceptions is conducted using the polygraph in most probation programmes. The examiner forms part of the containment team, providing probation officers and therapists with information about how honest the offender was when asked these questions. As with the usual lie detection test, the examiner ostensibly gives a verdict on the offender's veracity in one of three categories: no deception indicated; deception indicated; or inconclusive. The team then determines how to proceed with treatment and management of the offender on this basis, though often the next step is to confront the offender with the results and attempt to extract further information from him.

As two key UK proponents of polygraph exams, Grubin and Madsen (Grubin and Madsen, 2006: 479), state, the polygraph 'provides clinicians with more reliable sexual histories, more complete and accurate offence descriptions, and a greater likelihood of identifying high-risk behaviours, enabling intervention to take place before re-offending occurs.' Identifying high-risk behaviours and planning interventions based on these investigations is characteristic of the contemporary approach to sexual offenders, as noted by English (2004: 1255): it is a 'case management and risk control approach that is individualized based on each offender's characteristics.' Understanding the specific desires, fantasies, motivations, etc. of offenders is a precursor to determining their exact risk profile, and the polygraph provides a means through which such a profile can be constructed.

Risk management mechanisms have come to be central techniques in contemporary justice. In their influential work on the 'new penology', Feeley and Simon (1992, 1994) argue that by the 1990s certain criminal processes had begun to be managed through statistically based 'actuarial justice', concerned with the control of risks posed by dangerous groups rather than the punishment of individual criminals. At the same time, public anger and demands for retribution produced some of the most punitive responses to crime in the twentieth century (Garland, 2001; Simon and Feeley, 2003). The containment approach sits well within both trends, for it emphasises risk management and is aligned with the sweeping changes to legislation in the USA and UK which singled-out sexual offenders for additional control and surveillance during probation.

The period in which the containment approach developed was rife with such heterogeneous innovations in justice, as 'new ways of thinking and doing [were] assembled together from existing intellectual, social and material resources, in response to perceived problems of the time' (O'Malley, 2004: 10). The adoption of the polygraph into US and UK probation programmes represents an example of such an assemblage. Despite the heterogeneity of these innovations, Rose (2000) argues that they are nonetheless coherent, when understood as responses to the problem of how to control deviations from normal behavior *in a 'free society'*. He discerns two general tendencies in this regard:

> those [strategies] that seek to regulate conduct by enmeshing individuals within circuits of inclusion and those that seek to act upon pathologies through managing a different set of circuits, circuits of exclusion.
>
> (Rose, 2000: 324)

On this account, sex offender containment fits the exclusionary trend in control strategies, which attempt to manage risky individuals and to 'neutralize the dangers they pose' (Rose, 2000: 330). The development of polygraph programmes for the management of sex offenders thus appears to align with Rose's broad outline. He also argues that paedophiles represent a group which is doubly excluded. Unlike other risky individuals who are brought into confinement until their riskiness can be managed, 'Sexual predators, paedophiles, the incorrigibly anti-social are representatives of a new "human kind" – individuals whose very make up as human beings appears somehow faulty or incomplete, and whose very nature thus seems to place them permanently beyond the limits of civility and its demands on subjectivity' (Rose, 2000: 333–334).

I agree that sex offenders are constructed as monsters and have been subject to greater degrees of exclusion, for example through the development of name-and-shame types of legislation (such as Megan's Law) or through their subjection to castration treatments. And there are clear cases in the containment approach literature which characterise paedophiles as 'somehow faulty or incomplete', making them permanently risky and 'beyond the limits of civility'. This is most explicit in how the containment approach constructs offenders as fundamentally different and incurable. The Colorado state guidelines are clear on this:

> Many offenders can learn through treatment to manage their sexual offending behaviors and decrease their risk of re-offense. Such behavioral management should not, however, be considered a "cure," and successful treatment cannot permanently eliminate the risk that sex offenders may repeat their offenses.[31]

Other US guidelines on the containment approach mirror this pattern – for example, Arizona guidelines state: 'Sexual offending is a behavioral disorder which can often be managed but not "cured".'[32] And the same can be found in the UK context, as Grubin (1998: 42) comments: 'Sex offending is not a disease, and it does not lend itself to a cure. . . . One can, however, lower the risk of such re-offending significantly in some individuals, and decrease the frequency of offending in others.'

However, as can also be discerned in these quotations, the use of the polygraph in probation services has facilitated the – at least partial – re-entry of most sex offenders into risk management techniques, rather than keeping them permanently sequestered. This allows them to live within the community rather than being indefinitely placed in detention.[33] The polygraph's role comes to be a kind of civilizing technique, helping conquer the denial and deceptions of sexual offenders and allowing them to take part in truthful talk.

As Smart (2009: 552) argues, truth

> has come to occupy a special place in personal life, or at least in how it is believed that personal life should be lived. It is perceived as the source of healing (of the self) and, normatively speaking, the only acceptable basis for 'proper' personal relationships.

Overcoming the deceptions, lies and secrets of sex offenders forms part of this broader privileging of truth in health and healing. It is through confession of deviance that the healthy self can be restored. Foucault (1978) traces this alliance between truth and health to the shifts in Christian forms of confession, especially as regards its emphasis on the body as the location of sin. The practice of confession, he argued, imposed meticulous rules of self-examination, increasingly tied to the sins of the flesh, constituting sexual desire as the secret root of all evil. As the secular sciences of the mind (psychology, psychiatry and psychotherapy) developed their own modes of confession, Foucault (1978: 67) argues that the excavation of one's secret fantasies, desires, intentions, dreams and so forth came to serve the aims of therapy:

> [T]he obtaining of the confession and its effects were recodified as therapeutic operations. Which meant first of all that the sexual domain was no longer accounted for simply by the notions of error or sin, excess or transgression, but was placed under the rule of the normal and the pathological. [. . .] Spoken in time, to the proper party, and by the person who was both the bearer of it and the one responsible for it, the truth healed.

This framework of confession, truth and healing partly shapes the ways in which sex offenders are managed and treated in the containment approach, for its primary measurement for success is explicitly the amount of 'clinically significant disclosures' (CSDs) made by offenders.

Such CSDs are statements of offenders made before, during or after the polygraph examination, which are understood to be relevant to their treatment or supervision. This might relate, for example, to previous criminal activities or any suspected current criminal activity. But it is also more generally about understanding their desires, fantasies, masturbation habits, sexual relations and so on. That such statements can occur at any time in the polygraph examination process indicates that here again it is not used as a 'lie detector' in any objective fashion, but rather as a confessional apparatus. The aim is to get the offender to speak the truth about his deviance to produce greater trust in his willingness to self-manage a risky identity.

In England, the pilot of the technology found that 864 disclosures were made when the polygraph was used to interview offenders on probation, compared with 378 when it was not. This was taken to be direct evidence of the efficacy of the technique (Gannon et al., 2012), irrespective of whether the polygraph was actually switched on or not when these disclosures were made. Offenders making CSDs are thus understood to be cooperating in their treatment in good faith. Whether or not the polygraph can detect lies or not is largely irrelevant, as one article makes clear:

> [T]he polygraph is not a test, but a treatment tool designed to elicit a client's admissions to past behaviors and monitor current behaviors. Many therapeutic interventions that do not meet the standards requiring adequate documentation of practice standardization, reliability, and validity are nonetheless effectively utilized in the field.
>
> (Ahlmeyer et al., 2000)

Through the integration of the polygraph into sex offender management, the ability to induce confessions irrespective of scientific validity has been reframed within therapeutic and risk discourses such that the polygraph machine becomes a 'treatment tool'. The polygraph machine is used to encourage offenders to provide CSDs to demonstrate that they are reliable, consistent subjects over the course of their treatment and probation. Each examination, generally every six months or so, requires offenders to show that they understand that they are fundamentally incurable but that they are nonetheless managing this deviant identity appropriately by performing it openly within the treatment and surveillance infrastructure. As with other sexual monsters, lie detection practices serve to facilitate the management of a truth problem, in which other more common mechanisms for establishing trust and assuming veracity fail. Moreover, with the sex offender, the polygraph as a confessional apparatus is combined with confession as a form of therapy to achieve the re-integration of an otherwise permanently excluded group back into the risk management assemblage.

Rose (2000) argues that whilst risk has become a central feature of inclusion and exclusion strategies, the local techniques implemented are rarely actuarial in practice, as Feeley and Simon (1992, 1994) had claimed of the 'new penology'. Instead, he argues, risk techniques tend to remain qualitative rather than quantitative in nature and risk-based judicial practices are generally better characterised by the notion of 'risk thinking' (Rose, 2000: 332) than they are by actuarial technique. The emphasis on managing the future risk posed by offenders exemplifies risk thinking, for it is certainly 'concerned with bringing possible future undesired events into calculations in the present, making their avoidance the central object of decision-making processes, and administering individuals, institutions, expertise and resources in the service of that ambition' (Rose, 2000: 332).

At the same time, there is a growing degree of quantitation in the management of paedophiles with polygraphy, and in this regard the polygraph has helped to move the containment approach and the UK system further towards a more properly actuarial mode of risk management. Moreover, research in the judgement of sex offender risk has begun to challenge the professional, clinical opinion of individual risk, with some studies arguing that it is only marginally better than chance (Hanson, 1998) and that formal, quantitative procedures for risk assessment are at least as good, if not better (Grove and Meehl, 1996). There are now several risk-assessment 'technologies', including the Rapid Risk Assessment for Sex Offender Recidivism (Hanson, 1997), which produces a numeric score, to screen offenders into relative risk levels. Another, the Structured Anchored Clinical Judgement (or SACJ), is a three-step process taking into consideration a broader range of risk factors (Grubin, 1998). In both systems points are accumulated for the presence of specific risk factors (e.g. relationship to victim, marital status, substance abuse issues). The polygraph supports such processes by producing CSDs which are required to calculate risk levels.

Since the 1990s, sex offenders, most especially paedophiles, have thereby undergone a shift in their construction, from being men who made bad choices and needed social correction, to men who are fundamentally deviant, who cannot be cured or trusted. The polygraph has played a crucial role as a therapeutic tool in the management and treatment of such men, for it facilitates the production of 'clinically significant disclosures' in ways which support a situated performance of authenticity. It enables medical and legal professionals to manage offenders through risk techniques by quantifying the degree to which these men are deviant and deceptive. At the same time, it helps to train offenders to change their relationship with secrecy, deception and denial, producing subjects whose behaviours are more in keeping with social expectations of truth. In this regard, the sex offender as monster demonstrates contiguity with the monsters who have come before him, but also divergence, for the medical pretence which examiners previously used to help extract confessions of judicial significance has now become reality. Polygraph examiners have become experts at the nexus of medicine and law, helping to sustain a complex and otherwise precarious mix of heterogeneous scientific and legal practices in the management of a monstrous sexual subject.

Conclusion

Some groups of people are constructed as being fundamentally deceptive, manipulative and secretive, or as being more vulnerable to the temptations of such sinful actions. Their authenticity and sincerity are placed into doubt by the subject positions which they must inhabit as they navigate their lives within broader relations of power. Women, communists, gays and sex offenders have all become subjects of suspicion. They are people whose veracity remains permanently in question.

In this chapter I have drawn attention to the ways in which Othering processes contribute to the production of such subjects and, reciprocally, how being a person whose veracity is discredited contributes to being Othered. Through these and other processes suspect subjects become part of moral panics: they come to embody broader fears of difference and deviance, often at times of significant socio-political change. It is partly because of their deceptive character that these subjects become monsters, for it is not only that particular groups are deemed to be fundamentally different to 'normal people', which makes them monsters, but also that their difference is not easily discerned. This monstrosity within communities is a product of these figures' ability to appear to be normal through use of secrets, lies, deceptions and so on. In this way, the conversion of subjects of suspicion into monsters reflects broader problems of truth and trust in a given period, evidenced by the problematisations of veracity in patriarchy, the Cold War and risk society.

A common outcome for these groups is for them to be subjected to greater scrutiny under legal, scientific and medical practices, which are often already implicated in the production of suspicion. Producing knowledge about people whose confessions or denials cannot be trusted is a key objective of these practices during moral panics. It is because of their gendered, sexual and different bodies that people often become monstrous, and so it is their bodies that also become subjected to lie detection practices. The body as the source of deception and secrecy paradoxically becomes the source of truth – although it has been enrolled into different lie detection practices in different ways over time. The body has always served as a surface upon which answers can be written to otherwise unresolvable questions.

When involved with sexual monsters, lie detection practices often work alongside more violent means of managing the socio-political problems which such monsters embody. Whether it is the burning and drowning of witches, the persecution and unfair dismissal of homosexuals or the name-and-shame surveillance of sex offenders, there is often a more punitive counterpart to the lie detector's apparent civility and objectivity. In this regard, taken in the context of the history of brutality and scrutiny directed at subjects of suspicion, the polygraph is generally a tool not of civility but of complicity.

Notes

1 New York State Division of Probation and Correctional Alternatives (2009) *Research Bulletin 3: The Use of the Polygraph in Sex Offender Management*, www.criminaljustice.ny.gov/opca/pdfs/sopolygraphresearchbulletin3.pdf, last accessed 07.07.2017.
2 Centre for Sex Offender Management (2007) *The Importance of Assessment in Sex Offender Management: An Overview of Key Principles and Practices*, www.csom.org/pubs/assessment_brief.pdf, last accessed 06.07.2017.
3 Tennessee House Bill Number HB0308, 1995.
4 Colorado Sex Offender Management Board (1996) *Standards and Guidelines for the Assessment, Evaluation, Treatment and Behavioral Monitoring of Adult Sex Offenders*, www.colorado.gov/pacific/dcj/node/169581, last accessed 06.07.2017.
5 For information on Hawai'i see Tanabe J. (2000–2001); for information on Florida, see Shaw J.M. (2002–2003); for Illinois see Blair W.W. (2004–2005); and for Kansas see Eastman K.C. (1998–1999).

6 California Coalition on Sexual Offending (2004) *Position Paper for Clinical Polygraph Examinations in Sex Offender Treatment*, https://ccoso.org/sites/default/files/polygraphdraft.pdf, last accessed 06.07.2017.
7 Joint Polygraph Committee on Offender Testing (1998) *Recommended Guidelines for Clinical Polygraph Examinations of Sex Offenders*, www.dshs.texas.gov/csot/csotpoly.doc, last accessed 06.07.2017.
8 Indiana Polygraph Association (2000) *Standards and Guidelines for Clinical Examination of Sex Offenders*, www.indianapolygraphassociation.com/testing.asp, last accessed 04.07.2017.
9 New Mexico Sex Offender Management Board (2004) *Research Overview: Post-Conviction Sex Offender Testing*, www.forensic-centre.com/files/PCSOTOverview.pdf, last accessed 04.07.2017.
10 South Carolina Department of Probation, Parole and Pardon Services (2006) *Sex Offender Management Supervision Strategies*, www.dppps.sc.gov/Offender-Supervision/Supervision-Strategies/Sex-Offender-Management, last accessed 04.07.2017.
11 Bureau of Justice Statistics (2016) *Criminal Victimization, 2015*, www.bjs.gov/content/pub/pdf/cv15_sum.pdf, last accessed 07.06.2017.
12 103rd Congress, House Bill 3355.
13 105th Congress, House Report 256.
14 For mention of this brief investigation in the 1960s see: Security Commission (1983) *Report of the Security Commission into the Case of Geoffrey Prime.*
15 Ibid.
16 Central Intelligence Agency, Federal Bureau of Investigation and National Security Agency.
17 See, Security Commission (1983) *Report of the Security Commission into the Case of Geoffrey Prime*; and see Hansard HL Deb 12 May 1983 vol. 442 cc699–702WA.
18 See Hansard HC Deb 19 March 1986 vol. 94 cc293–9.
19 HC Deb 19 March 1986 vol. 94 cc293–9
20 Council of Civil Service Unions and others v. Minister for the Civil Service, House of Lords [1985] 1 AC 374, [1984] 3 All ER 935, [1984] 3 WLR 1174, [1985] ICR 14, [1985] IRLR 28, (44 CLJ 1), (101 LQR 153, 180).
21 For an overview of how the story evolved, see the following articles from *The Guardian*: 'Judge defends workers' rights over GCHQ ban', 17 July 1984; 'Lie detector tests endanger the innocent, say psychologists', 4 August 1984; and 'Thatcher "obliged to be fair on GCHQ"/House of Lords hears appeal over union ban at Cheltenham intelligence centre', 9 October 1984.
22 See, for example, New Scientist 'Polygraphs Condemned for Branding the Innocent', 15 December 1983, p. 795; and New Scientist 'How Cheltenham entered America's back yard', 5 April 1984, p. 8.
23 British Psychological Society (1986) Report of the Working Group on the use of the Polygraph in Criminal Investigations and Personnel Screening, *Bulletin of the British Psychological Society*, 39: 81–94.
24 S. 227 CJA 2003.
25 These are defined in section 28(4) of the 2007 Act.
26 See also The Polygraph Rules 2009 SI 619/2009 which came into force at the same time for the same purpose.
27 Travis, A. (2013) Lie Detector Tests to be Introduced to Monitor Sex Offenders, *The Guardian*, 1 July 2013, www.theguardian.com/uk/2013/jul/01/lie-detector-tests-sex-offenders, last accessed 06.07.2017.
28 British Psychological Society Working Party (2004) *A Review of the Current Scientific Status and Fields of Application of Polygraphic Deception Detection*, www.bps.

org.uk/content/review-current-scientific-status-and-fields-application-polygraphic-deception-detection, last accessed 10.07.2017.

29 Deschutes County Sex Offender Conditions, www.deschutes.org/justice/page/sex-offender-conditions, last accessed 10.07.2017.

30 Administrative Office of the Courts of Arizona (1999) *Arizona Standards and Guidelines for the Effective Management of Adult Sex Offenders on Probation*, https://antipolygraph.org/documents/az-sex-offenders.pdf, last accessed 07.07.2017.

31 Colorado Sex Offender Management Board (1996) *Standards and Guidelines for the Assessment, Evaluation, Treatment and Behavioral Monitoring of Adult Sex Offenders*, www.colorado.gov/pacific/dcj/node/169581, last accessed 06.07.2017.

32 Administrative Office of the Courts of Arizona (1999) *Arizona Standards and Guidelines for the Effective Management of Adult Sex Offenders on Probation*, https://antipolygraph.org/documents/az-sex-offenders.pdf, last accessed 07.07.2017.

33 Although they will always remain branded through public notification laws, and the CJA 2003 does allow for lifelong prison sentences for sex offenders where they are deemed to be too risky to be allowed back into the community.

References

Ahlmeyer, S., Heil, P., Mckee, B. and English, K. (2000) The Impact of Polygraphy on Admissions of Victims and Offenses in Adult Sexual Offenders, *Sexual Abuse: A Journal of Research and Treatment* 12 (2): 123–138.

Alder, K. (1998) To Tell the Truth: The Polygraph Exam and the Marketing of American Expertise, *Historical Reflections* 24 (3): 487–525.

Arthurs, A. (1985) GCHQ and the Changing Face of Staff Relations in the Civil Service, *Industrial Relations Journal* 16 (2): 26–33.

Bauman, Z. (2004) *Identity: Coversations with Benedetto Vecchi*, Cambridge: Polity Press.

Bayer, R. (1981) *Homosexuality and American Psychiatry: The Politics of Diagnosis*, Princeton, NJ: Princeton University Press.

Becker, J.V. and Quinsey, V.L. (1993) Assessing Suspected Child Molesters, *Child Abuse and Neglect* 17 (1): 169–174.

Beech, A.R., Fisher, D.D. and Thornton, D. (2003) Risk Assessment of Sex Offenders, *Professional Psychology: Research and Practice* 34 (4): 339–352.

Blair, W.W. (2004–2005) The Illinois Sexually Dangerous Persons Act: The Civilly Committed and Their Fifth Amendment Rights, or Lack Thereof, *Southern Illinois University Law Journal* 29: 461–480.

Bunn, G.C. (2012) *The Truth Machine: A Social History of the Lie Detector*, Baltimore, MD: Johns Hopkins University Press.

Chapman, R.D. (2008) A Review Of: "John F. Sullivan: Gatekeeper: Memoirs of a CIA Polygraph Examiner", *International Journal of Intelligence and Counterintelligence* 21 (2): 401–405.

Cohen, S. (2002) *Folk Devils and Moral Panics: The Creation of the Mods and Rockers* (3rd Edition), London: Routledge.

Cook, R.W. and Robinson, J.C. (1947) *Continued and Extended Use of the Polygraph, Department of Energy Declassification Review*, http://antipolygraph.org/documents/RHTG40422.shtml, last accessed 24/07/2017.

Cossins, A. (2015) *Female Criminality: Infanticide, Moral Panics and the Female Body*, Basingstoke: Palgrave Macmillan.

Critcher, C. (2002) Media, Government and Moral Panic: The Politics of Paedophilia in Britain 2000–1, *Journalism Studies* 3 (4): 521–535.
Critcher, C. (2008) Moral Panic Analysis: Past, Present and Future, *Sociology Compass* 2 (4): 1127–1144.
Daston, L. and Galison, P. (1992) The Image of Objectivity, *Representations* 40 (0): 81–128.
Denery II, D.G. (2015) *The Devil Wins: A History of Lying from the Garden of Eden to the Enlightenment*, Princeton, NJ: Princeton University Press.
Eastman, K.C. (1998–1999) Sexual Abuse Treatment in Kansas's Prisons: Compelling Inmates to Admit Guilt, *Washburn Law Journal* 38: 949–976.
English, K. (2004) The Containment Approach to Managing Sex Offenders, *Seton Hall Law Review* 34: 1255–1272.
English, K., Heil, P. and Veeder, G. (2016) The Containment Approach: A Strategy for the Community Management of Sex Offenders, pp. 713–743 in A. Phenix and H. Hoberman (eds.) *Sexual Offending*, New York, NY: Springer.
English, K., Pullen, S.K. and Jones, L. (1997) *Managing Adult Sex Offenders in the Community: A Containment Approach*, Washington, DC: US Department of Justice.
Fahs, B., Dudy, M. and Stage, S. (2013) *The Moral Panics of Sexuality*, Basingstoke: Palgrave Macmillan.
Feeley, M.M. and Simon, J. (1992) The New Penology: Notes on the Emerging Strategy of Corrections and Its Implications, *Criminology* 30 (4): 449–474.
Feeley, M.M. and Simon, J. (1994) Actuarial Justice: The Emerging New Criminal Law, pp. 173–201 in D. Nelken (ed.) *The Futures of Criminology*, London: Sage.
Fejes, F. (2016) *Gay Rights and Moral Panic: The Origins of America's Debate on Homosexuality*, Basingstoke: Palgrave Macmillan.
Foucault, M. (1978) *The History of Sexuality: An Introduction* (Vol. 1), New York, NY: Pantheon Books.
Foucault, M. (2003) *Abnormal: Lectures at the Collège De France 1974–1975*, London: Verso.
Fuchs, E. (1985) Who Is Hiding the Truth? Deceptive Women and Biblical Androcentrism, pp. 137–144 in A.Y. Collins (ed.) *Feminist Perspectives on Biblical Scholarship*, Chico, CA: Scholars.
Furby, L., Weinrott, M.R. and Blackshaw, L. (1989) Sex Offender Recidivism: A Review, *Psychological Bulletin* 105 (1): 3–30.
Gannon, T.A., Wood, J., Pina, A., Vasquez, E. and Fraser, I. (2012) *The Evaluation of the Mandatory Polygraph Pilot*, Ministry of Justice, www.gov.uk/government/uploads/system/uploads/attachment_data/file/217436/evaluation-of-mandatory-polygraph-pilot.pdf, last accessed 07/07/17.
Garland, D. (2001) *The Culture of Control: Crime and Social Order in Contemporary Society*, Oxford: Oxford University Press.
Geis, G. and Bunn, I. (1997) *A Trial of Witches: A Seventeenth-Century Witchcraft Prosecution*, London: Routledge.
Greenfeld, L.A. (1997) *Sex Offenses and Offenders: An Analysis of Data on Rape and Sexual Assault*, Washington, DC: US Department of Justice.
Grove, W.M. and Meehl, P.E. (1996) Comparative Efficiency of Informal (Subjective, Impressionistic) and Formal (Mechanical, Algorithmic) Prediction Procedures: The Clinical–Statistical Controversy, *Psychology, Public Policy, and Law* 2 (2): 293–323.
Grubin, D. (1998) *Sex Offending against Children: Understanding the Risk*, Police Research Series Paper 99, London: Home Office.

Grubin, D. and Madsen, L. (2006) Accuracy and Utility of Post-Conviction Polygraph Testing of Sex Offenders, *The British Journal of Psychiatry* 188 (5): 479–483.
Hanson, R.K. (1997) *The Development of a Brief Actuarial Risk Scale for Sexual Offense Recidivism*, Ottawa: Department of the Solicitor General of Canada.
Hanson, R.K. (1998) What Do We Know About Sex Offender Risk Assessment?, *Psychology, Public Policy, and Law* 4 (1–2): 50–72.
Johnson, D.K. (2004) *The Lavender Scare: The Cold War Persecution of Gays and Lesbians in the Federal Government*, Chicago, IL: University of Chicago Press.
Leo, R.A. (2008) *Police Interrogation and American Justice*, Boston, MA: Harvard University Press.
Littlefield, M.M. (2011) *The Lying Brain: Lie Detection in Science and Science Fiction*, Ann Arbor: University of Michigan Press.
McCoy, A. (2007) *A Question of Torture: CIA Interrogation, from the Cold War to the War on Terror*, New York, NY: Owl Books.
McGrath, R.J., Cumming, G.F., Burchard, B.L., Zeoli, S. and Ellerby, L. (2010) *Current Practices and Emerging Trends in Sexual Abuser Management*, Brandon, VT: Safer Society Press.
Merchant, C. (2008) Secrets of Nature: The Bacon Debates Revisited, *Journal of the History of Ideas* 69 (1): 147–162.
Neuilly, M.-A. and Zgoba, K. (2006) Assessing the Possibility of a Pedophilia Panic and Contagion Effect between France and the United States, *Victims and Offenders* 1 (3): 225–254.
Newburn, T. (2007) "Tough on Crime": Penal Policy in England and Wales, *Crime and Justice* 36 (1): 425–470.
O'Malley, P. (2004) *Risk, Uncertainty and Government*, Oxon: Routledge.
Park, K. (2006) *Secrets of Women: Gender, Generation, and the Origins of Human Dissection*, New York, NY: Zone Books.
Pratt, J. (2002) Critical Criminology and the Punitive Society: Some New "Visions of Social Control", pp. 168–184 in K. Carrington and R. Hogg (eds.) *Critical Criminology: Issues, Debates, Challenges*, Cullompton: Willan Publishing.
Rose, N. (2000) Government and Control, *British Journal of Criminology* 40 (2): 321–339.
Rowe, L. and Cavender, G. (1991) Caldrons Bubble, Satan's Trouble, but Witches Are Okay: Media Constructions of Satanism and Witchcraft, pp. 263–275 in J. Richardson, J. Best and D. Bromley (eds.) *The Satanism Scare*, New York, NY: Aldine de Gruyter.
Sharp, J. (2000) Refiguring Geopolitics: The Reader's Digest and Popular Geographies of Danger at the End of the Cold War, pp. 332–352 in K. Dodds and D. Atkinson (eds.) *Geopolitical Traditions: A Century of Geopolitical Thought*, London: Routledge.
Shaw, J.M. (2002–2003) Casenote: Sex, Lies and Polygraph Machines: The Portrait of Mr. Cassamassima, *University of Miami Law Review* 57: 429–474.
Simon, J. and Feeley, M. (2003) The Form and Limits of the New Penology, pp. 75–116 in T.G. Bloomberg and S. Cohen (eds.) *Punishment and Social Control* (2nd Edition), New York, NY: Aldine de Gruyter.
Smart, C. (2009) Family Secrets: Law and Understandings of Openness in Everyday Relationships, *Journal of Social Policy* 38 (4): 551–567.

Stokes, L.P. (2011) *Demons of Urban Reform: Early European Witch Trials and Criminal Justice, 1430–1530*, Basingstoke: Palgrave Macmillan.
Sullivan, J.F. (2007) *Gatekeeper: Memoirs of a CIA Polygraph Examiner*, Washington, DC: Potomac Books.
Tanabe, J. (2000–2001) Right against Self-Incrimination v. Public Safety: Does Hawai'i's Sex Offender Treatment Program Violate the Fifth Amendment, *University of Hawai'i Law Review* 23: 825–856.
West, D. (2000) Paedophilia: Plague or Panic?, *The Journal of Forensic Psychiatry* 11 (3): 511–531.
Young, J. (2003) Winning the Fight against Crime? New Labour, Populism and Lost Opportunities, pp. 33–47 in R. Matthews and J. Young (eds.) *The New Politics of Crime and Punishment*, Abingdon: Routledge.

Chapter 8

Lying

> They had me believing I was guilty. They had me believing I did something I didn't do. As soon as I got into the fresh air, I said I didn't do this but they said "Well you already confessed. It doesn't matter anymore."[1]

Daniel Gristwood, found guilty of second-degree attempted murder of his wife, Christina, had spent nine years in prison when Mastho Davis, a man with a life-long career of incarcerations for various felonies and misdemeanours, walked into a police station and confessed to the crime. But Gristwood had also confessed, after being awake for thirty-four hours, following a night out drinking. He had been interrogated for fifteen hours in a small cell, given limited food and drink, and endured a four-hour-long polygraph exam. The police lied to him and bluffed their way into a confession, telling him that the machine showed he was in denial about having tried to kill his wife. Although some disregarded Mastho Davis' declaration of guilt, believing that they had put the right man behind bars, it now seemed to others that Daniel Gristwood might have given a false confession, and his lawyer duly sought an appeal.

In 2003 Davis voluntarily entered the New York State Police Barracks to give statements regarding an attack he had committed some time ago, on a woman in her home, using a claw hammer. His description of the crime scene and the attack matched Christina's brutal assault exactly. In fact, there was a clear pattern to the crime that fit other assaults for which he had been convicted. First in March 1996, just two months after Christina's assault, he had entered a woman's home whilst she slept in bed with two children at her side. He beat her, choked her and threatened to murder her. He was caught and served five years in prison. Months after being released he committed a similar crime and was caught once again. During the second sentencing, Davis tried to confess to the Gristwood assault before a county court. The judge advised him to take counsel and to avoid further incriminating himself. After his second confession to a legal representative, and with new information to hand, it was up to the State Police as to whether they would take this most recent of Davis' confessions seriously, against the apparently polygraph-endorsed confession of Gristwood.

As they conducted their investigations, the police discovered that Davis was known in Syracuse for carrying a hammer around with him for protection. It also turned out that he had been sleeping in a family friend's apartment just two blocks from the Gristwood residence on the night of the attack. With fourteen prior arrests for assault, there was clear enough evidence that Davis had the opportunity to have committed the bludgeoning, that it was consistent with his behaviour in other attacks and that his confession was more than detailed enough to substantiate his presence at the scene of the crime. He was able to provide specific information about entering the house and of his actions in striking Christina. Mastho Davis looked to be the real assailant. Two years later, on 16 June 2004, Gristwood's conviction was vacated.

The polygraph was at the centre of the case. Just prior to his confession Daniel was subjected to four hours of testing. During the exam, his concentration lapsed and he repeatedly had to stop himself falling asleep from exhaustion; he had been drinking earlier that day and had been awake since the morning of the day prior. He had severe heart pains during the exam. Even the examiner was cautious about the results, reporting that the test was inconclusive and that it was likely affected by Gristwood's condition. But he also gave the police what they wanted, saying it appeared as if Gristwood might not be telling the whole truth. Returning to the interview with a more aggressive and deceptive line of questioning, Investigator Jerome set about trying to convince Gristwood to confess by claiming that the polygraph evidence demonstrated he was guilty. This kind of bluff or lie is not uncommon in polygraph interrogations, as I described in Chapter 6. The result of this harder stance was a confession from Daniel, retracted almost immediately, which nonetheless secured a conviction and put him behind bars for nine years. All the while Christina's assailant had remained at large. Not long after his release from prison, and only a few months after receiving damages for wrongful imprisonment, Daniel Gristwood died, bringing a tragic end to an awful story.

Lie detection

The cases of Daniel Gristwood and Peter Reilly demonstrate the ways in which lying in police practices, when combined with the lies and tricks involved in polygraph interrogations, can produce false confessions and wrongful imprisonment. Rather than closing the door on the history of torture in the West, the polygraph sits alongside it. Throughout this book, I have sought to account for the ways in which the lie detection practices of torture and polygraphy have been organised over time, and dealt with in different legal situations. How the cauldron was used to burn the arm of a suspect in trial by ordeal bears resemblance to the ways in which the blood pressure cuff is used in polygraphy. But there are also important differences. Certainly, they are of a different order with regard to physical duress, and their explicit logics are quite contrasting, one dependent upon God, the other on psychology. But I have shown how each helps to enact certain relations between the body and mind, truth and lies, in response to an epistemological failure. These technologies are used when other means of arriving at the truth have not worked, whether in the medieval application of ordeal to adjudicate claims of sexual assault, or in the contemporary use of the polygraph to manage and treat sex offenders. The polygraph is only the most recent recourse to the body, in a history of socio-legal problems of order and truth. Having long invoked torture as a curative factor, it is now impossible to separate it from the technologies upon which we draw today. Indeed, the third degree framed and helped to justify the polygraph's introduction (against the background of the third degree), expansion (alongside the CIA's experiments in physical and psychological brutality) and complex relations with the law (for example, in how *Miranda* and the Reid technique were entwined). Torture and technology are entangled as social institutions seek to respond to the political problems of truth and trust in any given era.

Lie detection depends upon a deceptive and manipulative performance, which further reveals how it is the quintessential expression of the socio-legal failure to derive the truth through honest means. When priests pretend to read burns as divine inscriptions or polygraphers pretend to interpret charts to indicate deception, they each deceive in order to bring an end to a seemingly irresolvable conflict of testimony. If the decline in state-sanctioned torture seemed to have augured the introduction of a more civil means of extracting confessions, it was not at the behest of truth, but at its continued expense. Embedding lie detection technologies into socio-legal practices does not expel deception, but means surreptitiously integrating or maintaining the use of deception in their enactment.

Superficially, the *Frye* ruling barred the polygraph from criminal trial, and the continued use of *Frye* in some states might be read to support an account of the polygraph as a technology firmly placed on the fringes of American justice. However, its success in policing and government has impacted upon trial decisions over the past century. Courts have had to make regular judgements on whether polygraph science and the professionalism of the polygraph community have improved to such a degree as might warrant admissibility. The courts have engaged in uncertainty work regarding its epistemological and ontological status. The outcome is that the polygraph has in fact found use in criminal trials in the United States: whether for short periods of experiment, such as in Massachusetts; in curious episodes of 'polygraph talk' found across the states; in the numerous courts which today allow admissibility with prior-stipulation; or in those select instances of longer-term admissibility, for example, in New Mexico. These positions demonstrate a complex and often uncertain set of rules and precedents in use at the State Supreme Court level, evidenced also in the growth of techniques available in the exclusionary toolkit to manage the ceaseless attempts to gain more stable access to the courtroom.

The work that scientists have done in changing how the polygraph works (developing software, adding new devices to the assemblage, creating new question techniques) and how their community is organised (creating new associations, training programmes, certifications and so forth) has resulted in numerous different implementations of 'the' polygraph machine across the United States. It is, in de Laet and Mol's (2000) terminology a 'mutable mobile', enacted differently at different times and in different locations. The legal mirror of the mutable mobile, evident in the polygraph's status in the US criminal justice system, is perhaps best characterised as a techno-legal hodgepodge, a mess of situated co-productions of technology, law and social order, which similarly resists classification and synthesis (Balmer, 2015).

Rather than being a death knell, then, the *Frye* decision and the subsequent development of this hodgepodge has contributed instead to the intensification of polygraph effects in some courtrooms. Lawyers, polygraph experts and police officers manipulate the uncertainties in the polygraph assemblage and in its legal status, in the police interrogation, in plea bargaining and in the courts. Rather than hamstringing the polygraph, this uncertain mess helps to hide its frequent role in criminal justice. The partial opacity and proximity afforded by this situation has the effect not of protecting defendants and juries from misleading claims of objectivity or usurpation of responsibility, as many courts' decisions have reasoned, but of masking the range of other techniques used by police officers and polygraph experts to deceive suspects into providing confessions of guilt. Since polygraph-induced confessions are readily admissible as evidence, but testimony to the polygraph's scientific status or situated use is often not, such confessions merely elide the danger of their production. Most worrisome is the manner in which the polygraph is used to cover over the police lies and manipulations which sometimes lead to false confessions and wrongful imprisonment.

Such troublesome outcomes are also a result of the role confession has played within relations of power, whether on the torture scaffold, in the trial by jury or in the polygraph examination. Scientists tend to dwell on the reduction of false positives in the development of lie detection technologies, in other words to reduce those cases in which an innocent person is incorrectly identified as a liar. However, the limit of the lie detector is not the false positive, though these are troubling misclassifications, but rather the false confession: this is the lie become truth. It, not silence, represents the point at which the power relations inherent to confessional modes of truth production become mere domination (Patton, 1989, 1994); the false confession pronounces the failure of a form of power in the false interrogation.

Lying and social order

The relations between lie detection and law present opportunities to reflect on lying more broadly, from a sociological perspective. As the polygraph has spread into a range of socio-legal spaces, it has variously interacted or integrated with, displaced or disrupted practices and their concomitant, situated conceptions of lying. In the developments and changes in torture and polygraphy we see how lying is embroiled in different arrangements of practices, shaped by but also supporting political and economic relationships invested in those practices.

For example, the polygraph integrated well with police interrogations and sex offender probation programmes, since these already depended upon confessional modes; but less well with the criminal jury trial, where it often failed to survive cross-examination and the exclusionary toolkit. Much of this is due to the way in which lying was already managed in such spaces.

Lying therefore, as much as truth, is part of knowledge-power relations, shifting over time as modes of knowledge production and justice are transformed. In this section, I return to philosophies of lying outlined in Chapter 1, reflect on what the case of lie detection might help us to understand about lying, and weave select sociological and political analyses into this account to explore the kinds of issues that a sociology of lying might develop.

Early philosophical explorations of lying are testament to the complexity of pinning down what a lie is, but also of how answers to the questions of lying embed political and social commitments. St Augustine (Schaff, 1887), for example, took his torch of truth to search the dark and winding caverns of lying in pursuit of answers to his questions regarding deceit. He asked, what is a lie, and could it ever be right to tell one? His answer to the latter was that the essential quality of lying was duplicity, a kind of double-speaking in which persons said something which they did not hold true in their mind. Since this moved a person's speech away from God and towards the devil, his answer to the second question was 'no', there is never a circumstance in which it is right to lie.

However, in some respects Augustine clouded the issues, especially regarding what constitutes a lie with regard to the intention to deceive. His analysis of the cynical acquaintance showed that one could tell the truth in order to deceive, or tell a falsehood in order to communicate the truth. In other words, Augustine failed to resolve whether lying required the intention to deceive or whether duplicity was sufficient. It was the intersubjective dimension which caused such troubles: how persons orient to another's speech, and how they then orient themselves in turn, seems to have introduced rather too much uncertainty regarding whether any given action could be counted as a lie or not. Retreating from this complexity, Augustine returned to the inner experience, for here it seemed he could rely on known intentions as guarantors of the truth, apparently free of the complications of intersubjective space. He reported two facts that any good Christian could hold on to. First, a certain lie is told when one speaks duplicitously, against one's thoughts, and with the intention to deceive. Second, it is a certain truth when one speaks in line with one's thoughts and has no intention to deceive. By only speaking all of what they truly held in mind and never meaning to deceive, Christians could be sure of never having told a lie. Since God is only true, reasoned Augustine, we should also be so (Griffiths, 2004). Augustine's investigations and struggles reveal long-standing uncertainties in lying and how it is that we manage truth and lies as part of everyday life. They also point towards the ways in which questions of knowledge, power and justice – in this case, each of these being powerfully shaped by Christian readings of biblical texts – become woven into considerations of lying, and vice-versa. For Augustine, there is a possible world in which we only ever speak the truth, knowledge serves the glory of God and justice is divine. Lying and deception, as seen in this account, often speak to questions regarding the nature of society.

In contrast to Augustine, Simmel (1950) argues that uncertainty and selectivity is ontologically fundamental to the very possibility of sociation: that we do not see the other in their totality, transparently and objectively, is a condition of our social experience and not one which we must necessarily try to overcome:

> Man's representation of man is thrown out of true by dislocations, additions and subtractions . . . which exert an a priori influence. . . . [A]s a matter of fact those changes and reshapings, which prevent this ideal recognition of him, are precisely the conditions through which the relationships which we know as the strictly social become possible.
>
> (Simmel, 1910: 380–381)

We live in a world of veiled impressions and generalisations of each other. In fact, Simmel (1950: 311–312) goes as far as to argue that if we were to try to communicate the whole truth about our thoughts, then we would be considered 'insane', for our inner experience is a 'chaotic whirling of images and ideas which objectively are entirely unrelated to one another'. As if overhearing this cacophony, he argues, we filter the 'usable' portions which are of relevance to whatever it is that we are up to. Nor is there any way around this:

> What is more, these fragments are not a representative selection, but one made from the standpoint of reason, value, and relation to the listener and his understanding. . . . With an instinct automatically preventing us from doing otherwise, we show nobody the course of our psychic processes. . . . Always, we show only a section of them, stylized by selection and arrangement.
>
> (Simmel, 1950: 312)

Sacks (1992: 549) puts things rather more succinctly by saying 'everyone has to lie.' For Sacks, one possible meaning of the word 'lying' is to designate when someone uses a permissible expression (i.e. it makes sense in the situation) but which is incorrect, given the state of affairs. His example is that when someone asks us 'How are you?' as a greeting, we cannot just say anything. We cannot say 'robots!' and be taken to have answered the question. It is not our intentions which guarantee the meanings of what we say. Rather, we have a choice from three subsets of words with which to respond. There are words which indicate a negative state (such as bad, rubbish, terrible); words which indicate a neutral state (okay, fine, well, etc.); and those which indicate a positive state (great, wonderful, fantastic, and so on). The consequence, Sacks argues, is that in some circumstances we are at least presented with the opportunity, and might in fact be required by the situation, to choose a word from a subset which does not match our state. For example, if an acquaintance from a long-forgotten party greets us by saying 'how are you?' as she passes us briskly on a busy street, we are not socially permitted to say, 'I'm terrible,' and then launch into a lengthy explanation of what is going on in our lives that makes things so very bad. Rather, we must say, 'I'm fine.' That this is recognisably the case shows that the combination of social norms with interactive sequences (like the greeting pair of phrases) means that we all must 'lie' in this way, in some circumstances.

In this regard, for Simmel and Sacks, Augustine's dream that we might exalt God's nature by always and only saying the whole truth of what we have in mind is not psychologically or socially possible, and perhaps far from desirable, for rather than supporting social order, the whole truth might in fact disrupt the very means through which order is maintained. Fibs and fabrications, and perhaps even outright lies and malicious deceptions, are a necessary part of everyday interaction to the extent that they support the ongoing orderliness of situations.

A sociological concern with the ontology of lying, then, must acknowledge its multiplicity. There is no single form of lying, for which we could provide a description which captures all instances. What we count as a lie, and whether a lie should count as an ethically problematic action, is determined contextually (Hunt and Manning, 1991) and resolved for the practical purposes in which actors are engaged. This is devastating for philosophical attempts to produce one single definition. But it is less problematic for sociologists, because it leads to further inquiry. Developing this insight, for example, a second sociological question of lying presents itself: how do people do the work of determining whether something counts as a lie and how do societies manage the uncertainty in lying, truth and intention produced by intersubjective experience? In this, the history of torture and lie detection technology has something to contribute. Before building towards that account, however, I wish first to introduce a further element of Simmel's analysis.

Lying and modernity

Simmel (1950) claims not only that we always filter our talk, but that society tends towards greater degrees of filtering, secrecy, lying and the myriad other forms of concealment and selection which arise in social discourse. Not because of a love of lies, or any greater degree of moral bankruptcy, as we might worry of our politicians today. Rather, he argues, it is because of modernity that we more selectively express ourselves in talk. In accordance with other social theorists, Simmel describes how the processes of modernisation (and all it entails of work, urban life and social relations) bring with it greater degrees of differentiation and individualisation, meaning we become less similar to each other over time. On this account, a consequence of living more closely to each other is that we are even less able to communicate the specificity of our experience to one another.

For truth, the consequences of the growing individualising forces of global consumerism are perhaps nowhere more clearly seen than in the massive expansion in the use of lie detection technology in the Cold War. As one of the most intense periods of international espionage, government secrecy and state surveillance, it is hard to imagine that the Cold War was not partly a response to problems of trust arising from an increasingly differentiated society, not only in the US but internationally. It also seems reasonable to assume that these practices reciprocally intensified a pattern of individualisation and selectivity by their effects on public and private discourse of the type which Simmel anticipates. Like Catholics practicing more secretly in their homes under Elizabethan threats of torture, some Americans under Cold War conditions found themselves less able to speak truthfully and were required to filter their talk more assiduously. But, as Protestants felt immune to these forces, so too would some thrive in an environment of greater despotism. Control and freedom over the truth are not dealt or felt equally through society.

Simmel goes further in his claims about the effects of modernity on lying, arguing that the shift towards a more differentiated society places a strain on trust as the mediator of social relations:

> Our modern life is based to a much larger extent than is usually realized upon the faith in the honesty of the other. Examples are our economy, which becomes more and more a credit economy, or our science, in which most scholars must use innumerable results of other scientists which they cannot examine. We base our gravest decisions on a complex system of conceptions, most of which presuppose the confidence that we will not be betrayed. Under modern conditions, the lie, therefore, becomes something much more devastating than it was earlier, something which questions the very foundations of our life.
>
> (Simmel, 1950: 313)

For him, it is not only that we lie more, but that lies become more destructive. In part, this is because of the proliferation of social forms in modernity (Simmel cites the modern economy and scientific forms of knowledge production, for example). In the late 1800s, he could not have anticipated the acceleration in technological diversity and its consequences for social relations, but his analysis does seem to apply. For instance, we might interpret the rise of social media to exemplify his claim. People now have available to them a whole range of ways of manipulating, deceiving and fabricating, which were simply not available at the turn of the twentieth century. Indeed, wholly fabricated, entire personas can be created through the click of a few buttons. At its extremes, this can become a dark obsession, for example in the well-known and seemingly pervasive social media phenomenon of 'cat-fishing'.

Further support for Simmel's analysis can be found in the ways in which new media technologies often spur concerns about their consequences for truth and lying. The recent growth of fake news websites and the claim that we now live in a post-truth society is just one such example. Although the idea of post-truth had been around for some time (see, for example, Keyes, 2004), journalists, politicians, scientists and a range of other experts began to flesh out the claim after surprising results in the UK 'Brexit' referendum in June 2016, and a few months later in the US presidential election. Social media was amongst the targets in the blame game over these shock political outcomes, because it seemed to many that the circulation of fake news stories across the networks of Facebook and Twitter had played a decisive role.

The case of 'post-truth society' also bears directly on the question of lie detection and lying. In response to this most recent of concerns regarding truth, computer scientists and other experts have sought to develop new forms of lie detection, for example using natural language processing (Rubin et al., 2015) to conduct linguistic and network-based analyses of news articles and social media (Conroy et al., 2015). New social forms breed new forms of lying, secrecy and deception, which result in new forms of lie detection and novel configurations of bodies, technology and truth. In this, there seems to be some evidence for Simmel's account that we not only lie more, but that these lies become more powerful and destructive in an era of greater individualisation (severing the UK from the EU, and electing Donald Trump, for instance).

But the history of torture and lie detection technology does not fully support his account because it evidences the long-standing complexity of lie detection forms and ontologies of lying. It even provides evidence for the role of technology. Returning to where I started this book, with the trial by ordeal, here we see some of Augustine's philosophy of lying put into practice. Since God is a living force in the medieval and early modern periods, we have to take seriously the idea that for those engaged in trials by ordeal God really did act as a kind of lie detector, speaking through seared flesh. Lying in the ordeal is conceptualised by reference to God's perfect relation to truth and the devil's status as the 'father of lies'. In this way, the trial by ordeal enacted a conception of lying within a broader religious ontology. What a lie is, and how we know one, made sense as part of these wider systems of making knowledge, resolving conflicts and dealing with the invisible crimes of the era. At the same time, however, historians have shown that a social mechanism was also in operation, as I discussed: the clergy used the time they spent engaged with suspects to determine if they were guilty of the crime and, if so, if they were adequately penitent. Ultimately, the clergy could take deliberate control of the outcome when it became necessary, since they judged the party innocent or guilty on inspection of the wounds. If a person had been wrongly brought before the ordeal (for example, by being a virtuous person but an outsider to a community, caught up in events beyond his/her knowledge), then the clergy could intervene through the practices of the ordeal to protect the innocent from torture or death. Entwined with a divine conception of truth and lying was a more mundane practice of determining veracity through face-to-face interaction and the everyday, mundane forms of judging whether to trust somebody. Yes, God spoke through the flesh, but it was not only his voice which shaped the ontology of lying and its practical determination. Lie detection helped produce an intimate interaction that made it possible for certain figures to trust their own judgement on the truth of a suspect's speech.

Thus, even the technological innovations which seem to support Simmel's claim are not without precedent. In trial by ordeal we already see technological practices side-by-side with social and legal practices, for the burning of the flesh with hot pokers, heated water or burning coals could easily be said to be a technical action (fire being perhaps one of humanity's most important technological innovations). On this account, the trial by ordeal was already a socio-technical co-production of lying at the nexus of religion, law and social order. The polygraph interrogation mirrors much of this organisation. Science acquires the authority which religion once held, the polygraph facilitates bodily inscription, both of which mask a more recognisably social process in which a kind of intimacy is rapidly produced. The mechanisms of bleeding, clearing and composition make for powerfully intimate disclosures from suspects to lie detection operators. These examiners then rely on their findings and their feelings to report to their policing colleagues on whether the subject seems trustworthy or not. The polygraph examiner is not quite a clergyman in modern dress, but there are important corollaries in practice.

As the ordeal and the polygraph show, there has long been a mixture of different socio-technical practices in operation for the judgement of veracity. These are combined, separated, adjusted and manipulated to produce what seem to those with the power to decide to be just or desirable outcomes. The polygraph's journey into the US criminal justice system, its complex status within the criminal trial court, use in the Cold War, adaptation to sex offender treatment programmes and gradual infiltration of the British criminal justice system are evidence of shifting social forms regarding truth – but perhaps only of the most recent changes in a much longer history, which does not easily conform to Simmel's broad-brush claims regarding modernity. They also show that if we have entered a post-truth era, some of its key qualities are not without precedent.

Lying, lie detection and uncertainty

Although philosophers and sociologists have opened up a number of avenues in relation to lying and its role in social order, what they have tended not to emphasise is the way in which certain groups become suspected of lying, or of being somehow different in regard to the lies that they tell, by frequency or by nature. In other words, we do not yet have an adequate picture of the biopolitics of lying. Taking a brief look at this issue points to ways in which uncertainty, lie detection and the embodied subject positions of certain social groups become woven into the fabric of lying.

The history of lie detection's relationship with gender and sexuality exemplifies the need to develop such an account. As outlined in Chapter 7, women have long been considered to be the deceptive sex. This is partly due to biblical teachings, but philosophy and literature too have played their part. As Denery

(2015) reports, biblical, philosophical and poetic works have long lamented that everybody lies, but in doing so have primarily focused on women's deceptions. A man's lies tend to be explained away by reference to the situation, which is seen in some way to have compelled him to lie. Plato's appraisal of Achilles' and Odysseus' lies in the Hippias Minor is instructive. For Hippias, what marks Achilles out as the better man is that when he lies his lies are compelled; he lies to ease other citizens' pain or fears. In contrast, women's lies become a reflection of who they are: men lie to get by in the world, but women are understood to be natural born liars. As Denery (2015: 203) puts it, Western thought assumes that 'when a woman lies she lies because she is a woman, and every woman is a liar.'

Clearly, this characterisation is unjust but there is some evidence for sex differences in lying (DePaulo et al., 1996). Under experimental conditions or in self-report, women seem to tell more other-oriented lies, for example, which are designed to help people save face or smooth social relations. But perhaps it is in light of patriarchal requirements that women regulate the feeling of the group, that it becomes more likely that women, rather than men, are compelled to lie by the situation. If this is the case, we might modify Sack's argument that 'everyone has to lie' with the phrase 'but some more than others'. In a system of deep inequality, the strategies adopted by those groups in subordinate, stigmatised or precarious positions may involve lying and secrecy to negotiate the relations of power in which their identity positions are implicated (Shilling and Mellor, 2015). LGBTQ individuals, for example, must often 'pass' as straight to gain access to sources of capital and everyday resources which others take for granted, and might even form groups akin to secret societies in homophobic social conditions (Ponse, 1976).

Such power relations become acute when the social actors' difference, and their attendant lies, secrets and performances become more powerfully troubling to social order. Witches, homosexuals, communists, sex offenders and Muslims have all become monstrous liars at one time or another, in part because of biopolitical forces. Whether it is urbanisation, population expansion and diversification, or war, the consequences of structural changes in the social order can precipitate the construction of monstrous liars, a process which is more or less intentionally designed to manage and contain these figures and the disruption which they cause, as can be seen in the history of moral panics.

However, the uncertainty of lying cannot be permanently contained, or fundamentally resolved. As Simmel points out, there is always selectivity in talk and unknowns remain even in our most intimate portraits of each other. Moreover, as Derrida (1997) contends in his deconstruction of truth and lies, even if we were able to demonstrate on some set of objective conditions that persons had not told the truth about their thoughts, there is always recourse – if only for individuals – to claim that they did not deceive, that there has been some mistake:

> for structural reasons, it will always be impossible to *prove*, in the strict sense, that someone has lied even if one can prove that he or she did not tell the

> truth. . . . [there will always be the possibility of] alleging the always possible difference between the said, the saying, and the meaning-to-say, the effects of language, rhetoric, and context: "I said that but that is not what I meant to say; in good faith, in my heart of hearts, that was not my intention; there has been a misunderstanding."
>
> (Derrida, 2002: 34)

Although divergent in their strategy and overall argument, what Derrida and Simmel share in common is an appreciation of the consequences for society writ large of the indeterminacy of understanding what the other has in mind. Although some part of much situated interaction serves to manage this uncertainty, it nonetheless arises again and again as a broader socio-political problem throughout history, bringing with it new attempts to develop modes of lie detection, new ways to know the intentions of others, and new faith in whatever discourse promises to bring an end to this uncertainty.

The jury, for example, developed as a specific legal form to deal with such trouble. Much like lie detection technologies, it has succeeded not because of its power to objectively clarify uncertainty, but because of its social force. The jury is the only lie detector which decides on guilt without recourse to deception. It is this which marks it out from the history of lie detection practices. Other modes, ostensibly, seek to dispel uncertainty by discerning truth at its source, in the body. This masks the underlying confessional power relations at the heart of these practices. In contrast, the jury takes up its role as lie detector precisely as a social determination of guilt where a confession might be absent, and does so explicitly. The bargain for its ability to do what other lie detectors cannot is that it should not be submitted to the cross-examination procedure. The jury is the black box which cannot be opened (Fisher, 1997). It is this compromise of the criminal jury trial which allows the jury's uncertain mechanism to overcome the uncertainty of lying over which it, in the end, is final arbiter.

By comparison, even the most successful lie detection techniques have failed. Rather than seeing an end to lies, deception and corruption in government and society, the utopian future promised by key figures in the development of the polygraph, we have witnessed a growth in lie detection without any concomitant decline in problems of truth and trust. Alongside his arguments regarding modernity, Simmel (1950) claims that secrecy and lying, and indeed secret societies, grow in proportion to the degree of control and domination in society: the greater the totalitarianism in social order, the greater the degree of concealment. Simmel stops short of developing this observation into a dialectical relationship, but we might venture, on the basis of the case of torture, lie detection and the law, that there is an ongoing, biopolitical struggle over veracity in the West, in which are implicated forces of control, modes of resistance (through concealment, deception and secrecy) and iterations of recourse to the body as a means of overcoming such resistance.

Arendt (1973) gets us closer to such a conclusion in her contention that the totalitarian forces of domination have given up on their long-held, terrible dream of mind reading and resorted instead to great projects of mapping, observing, recording and connecting information about the population. Failing to conquer lying, the state, she claims, resolved instead to conquer opacity and uncertainty more broadly. The emergence of risk management techniques (which disavow full knowledge in acceptance of uncertainties which can be prepared for) and the broad array of surveillance devices now in service to this aim of transparency (which, in compiling such vast quantities of data obscure as much as they do reveal) are testament to the dialectical struggle for mastery over the uncertainty of intention by reference to the body. What the family of monstrous liars evidences in this regard is that epistemological uncertainty can become attached to specific social figures in moments of trouble for social order. It is often those figures who stand in ontological opposition to (that is, those Others used to construct through difference) the heteronormative, patriarchal, white modes of identity central to social organisation in the West, who are most readily positioned as subjects of suspicion. When lying is problematised, it is often the figure of the monstrous liar who becomes the embodiment of fears and tensions and whose body, in the end, leads towards a temporary resolution of the problem.

Torture and lie detection technologies find homes in these uncertain times. The grey area which is 'the lie' is a source of uncertainty which at times recedes and is managed through day-to-day affairs, but at others comes to the fore and is threatening. It is the lie's indeterminacy which powers the dark potential of torture, and which is papered over in the seeming civility and promise of lie detection technologies.

Note

1 Daniel Gristwood, quoted in Green, P. (2005) Man Proclaims Innocence, *Post Standard*, Tuesday, 4 October.

References

Arendt, H. (1973) *The Origins of Totalitarianism*, London: Harcourt, Brace and Company.

Balmer, A.S. (2015) Telling Tales: Some Episodes from the Multiple Lives of the Polygraph Machine, in M. Pickersgill and E. Cloatre (eds.) *Knowledge, Technology and Society: Interrogating the Nexus*, Abingdon: Routledge.

Conroy, N.J., Rubin, V.L. and Chen, Y. (2015) Automatic Deception Detection: Methods for Finding Fake News, *Proceedings of the Association for Information Science and Technology* 52 (1): 1–4.

de Laet, M. and Mol, A. (2000) The Zimbabwe Bush Pump Mechanics of a Fluid Technology, *Social Studies of Science* 30 (2): 225–263.

Denery II, D.G. (2015) *The Devil Wins: A History of Lying from the Garden of Eden to the Enlightenment*, Princeton, NJ: Princeton University Press.

DePaulo, B.M., Kashy, D.A., Kirkendol, S.E., Wyer, M.M. and Epstein, J.A. (1996) Lying in Everyday Life, *Journal of Personality and Social Psychology* 70 (5): 979–995.

Derrida, J. (1997) History of the Lie: Prolegomena, *Graduate Faculty Philosophy Journal* 19 (2/1): 129–161.

Derrida, J. (2002) *Without Alibi*, Stanford, CA: Stanford University Press.

Fisher, G. (1997) The Jury's Rise as Lie Detector, *Yale Law Journal* 107 (3): 575–713.

Griffiths, P.J. (2004) *Lying: An Augustinian Theology of Duplicity*, Grand Rapids, MI: Brazos Press.

Hunt, J. and Manning, P. (1991) The Social Context of Police Lying, *Symbolic Interaction* 14 (1): 51–70.

Keyes, R. (2004) *The Post-Truth Era: Dishonesty and Deception in Contemporary Life*, New York, NY: St. Martin's Press.

Patton, P. (1989) Taylor and Foucault on Power and Freedom, *Political Studies* 37 (2): 260–276.

Patton, P. (1994) Foucault's Subject of Power, *Political Theory Newsletter* 6 (1): 60–71.

Ponse, B. (1976) Secrecy in the Lesbian World, *Urban Life* 5 (3): 313–338.

Rubin, V.L., Chen, Y. and Conroy, N.J. (2015) Deception detection for news: Three types of fakes, *Proceedings of the Association for Information Science and Technology* 52 (1): 1–4.

Sacks, H. (1992) Everyone Has to Lie, pp. 549–566 in E.A. Schegloff (ed.) *Lectures on Conversation*, Oxford: Blackwell.

Schaff, P. (1887) *A Select Library of the Nicene and Post-Nicene Fathers of the Christian Church, Volume 1: The Confessions and Letters of St. Augustin*, Buffalo, NY: The Christian Literature Company.

Shilling, C. and Mellor, P.A. (2015) For a Sociology of Deceit: Doubled Identities, Interested Actions and Situational Logics of Opportunity, *Sociology* 49 (4): 607–623.

Simmel, G. (1910) How Is Society Possible? *American Journal of Sociology* 16 (3): 372–391.

Simmel, G. (1950) *The Sociology of Georg Simmel*, New York, NY: Free Press.

Appendices

Appendix I

Source materials used for case of Daniel Gristwood

Court case opinions

- *Daniel G. Gristwood v. State of New York*, 39 Misc. 3d 1226(A), 971 N.Y.S.2d 71 (2013).
- *Daniel G. Gristwood v. State of New York*, 119 A.D.3d 1414, 990 N.Y.S.2d 386 (2014).
- *Gristwood v. The State of New York*, #2011009109, Claim No. 114040. (NY. Court of Claims, 2011)

Miscellaneous

- Daniel G. Gristwood, signed statement before investigator Frank J. Jerome, Town of Clay, County of Onondaga, State of New York, taken on 1 December 1996.
- *People of the State of New York v. Mastho Davis*, Onondaga County Courthouse, transcript, dated 6.18.2004.

News media

- *Justicedenied.org* (15 January 2015) Daniel Gristwood Dies 4 Months After New York Pays $7.3 Million for 9 Years Wrongful Imprisonment, http://justicedenied.org/wordpress/archives/2912, last accessed 10.09.2017.
- *Syracuse.com* (21 May 2014) Appellate Judges: State's Argument Against Dan Gristwood's Innocence "Doesn't make sense.", www.syracuse.com/news/index.ssf/2014/05/appeals_court_states_argument_against_dan_gristwoods_innocence_doesnt_make_sense.html, last accessed 10.09.2017.
- *Syracuse.com* (16 October 2011) D.A. William Fitzpatrick Responds to Assertions in the Case of Daniel Gristwood, http://blog.syracuse.com/opinion/2011/10/fitzpatrick_gristwood.html, last accessed 10.09.2017.
- *Syracuse.com* (9 October 2011) An 'Open and Shut Case' vs. an Innocent Man, www.syracuse.com/news/index.ssf/2011/10/an_open_and_shut_case_against.html, last accessed 10.09.2017.

- *The Post Standard* (14 April 2007) Syracuse Man Charged in Florida Attack; Mastho Davis Had Confessed to 1996 Clay Incident but Wasn't Prosecuted.
- *The Post Standard* (8 July 2006) Murder-try Case Dropped.
- *The Post Standard* (4 October 2005) Man Proclaims Innocence: Convicted of Attack on Wife, He's Freed after Other Inmate Confesses.
- *The Post Standard* (18 June 2004) Man Takes Blame in 1996 Claw Hammer Attack.
- *The Post Standard* (24 May 2004) Convict Says He Should be Freed.
- *Syracuse Herald Journal* (27 September 1996) Hammer Attacker Sentenced.
- *The Post Standard* (21 August 1996) Man Guilty of Murder Attempt.
- *Syracuse Herald Journal* (17 August 1996) Chemist Might be Allowed to Testify in Gristwood Case.
- *The Post Standard* (16 August 1996) Police Defend Suspect's Questioning.
- *The Post Standard* (15 August 1996) Police Say Accused Attacker Was Calm.
- *Syracuse Herald Journal* (14 August 1996) DA: Man Didn't Ask About Wife.
- *Syracuse Herald Journal* (4 March 1996) Husband Arraigned for Hammer Attack on Wife.
- *Syracuse Herald Journal* (24 January 1996) Mental Tests Set in Attack on Wife.
- *The Post Standard* (13 January 1996) Husband Accused of Trying to Murder Mother of Five.

Appendix 2

Source materials used for case of Peter Reilly

Court case opinions

- *Peter A. Reilly v. Edward P. Leonard, Individually and as Commissioner of the Connecticut State Police, Thomas J. McDonnell, Individually and as an officer of the Connecticut State Police Department*, 459 F. Supp. 291 (1978).
- *Peter A. Reilly v. State of Connecticut*, 32 Conn.Supp. 349, 355 A.2d 324 (1976).

Miscellaneous

- Transcript of polygraph examination/ interrogation of Peter Reilly.
- Connery, D. (1977) *Guilty Until Proven Innocent*, New York, NY: Putnam.
- Barthel, J. (1976) *A Death in Canaan*, New York, NY: E.P. Dutton.
- Freedom of Information Commission of the State of Connecticut Report in the Matter of a Complaint by Ruth Epstein and The Lakeville Journal Company, LLC. Against State of Connecticut, Department of Public Safety, Division of State Police. Docket #FIC 2003–320.
- Freedom of Information Commission of the State of Connecticut Report in the Matter of a Complaint by Donald S. Connery against the State of Connecticut, Department of Public Safety, Division of State Police. Docket #FIC 2003–313.

News media

- *The Lakeville Journal* (3 October 2013) Forty Years Later, and Still No Closure.
- *The Lakeville Journal* (16 October 2008) Barbara Gibbons and Peter Reilly: 35 Years after that Tragic Night.
- *The Lakeville Journal* (28 June 2007) Sponzo, Key Judge in Reilly Case, Dead at 92.
- *Litchfield County Times* (30 January 2004) In the Peter Reilly Case, Odd New Developments.

- *The Washington Post* (20 January 1979) Witness Steps Forward in Reilly Case.
- *New York Times* (19 January 1979) New Witness Found in 1973 Reilly Case.
- *New York Times* (18 January 1979) Police Checking into New Leads in Gibbons Case.
- *Wall Street Journal* (7 November 1977) A Graphic, True Detective Story.
- *New York Times* (12 June 1977) Reilly Report Highlights.
- *New York Times* (2nd June 1977) Connecticut Report Is Critical of Reilly's Prosecution.
- *Journal Tribune, Biddeford* (2nd June 1977) Death Case Authorities Are Cleared of Criminal Charges.
- *The Hour* (31 May 1977) Attorney Interviews Polygraph Expert in Reilly Investigation.
- *New York Times* (24 December 1976) One-Man Jury to Study Reilly Case Is Named by Judge in Connecticut.
- *The Washington Post* (26 March 1976) New Trial Is Ordered in Slaying.
- *New York Times* (19 February 1976) Hearing Is Ended on Reilly's Bid for a New Trial.
- *New York Times* (14 February 1976) Parmalee Brothers Deny Role in Gibbons Murder.
- *New York Times* (31 January 1976) Judge Says Lawyers for Reilly Have Met 2 of the 3 Prerequisites for a New Trial.
- *New York Times* (21 January 1976) Lawyers for Youth Convicted of Killing Mother Introduce Prints of a Neighbor.
- *New York Times* (15 December 1975) Arthur Miller Turns Detective in Murder.
- *New York Times* (16 December 1975) Arthur Miller and Others Contend Clock Absolves Youth Convicted of Matricide.
- *New York Times* (25 May 1974) Youth Sentenced in Mother's Death.
- *New York Times* (13 April 1974) Youth Convicted of Killing Mother.

Index

For Product Safety Concerns and Information please contact our EU
representative GPSR@taylorandfrancis.com
Taylor & Francis Verlag GmbH, Kaufingerstraße 24, 80331 München, Germany

www.ingramcontent.com/pod-product-compliance
Ingram Content Group UK Ltd.
Pitfield, Milton Keynes, MK11 3LW, UK
UKHW022125190926
475702UK00013B/1659
* 9 7 8 0 3 6 7 8 7 2 5 0 2 *